高等院校建筑·艺术设计专业教学改革丛书
Teaching Reform Series on Architecture & Art Design in University

表面的深度
The Depth of the Surface

中法联合教学教案——绘画·空间·设计

(法) Philippe Guérin
法国巴黎玛拉盖国立高等建筑学院
法国诺曼底国立高等建筑学院
中国东南大学建筑学院 编著
隽石(中国)文化传播机构
JUNSTONE 策划
东南大学出版社

图书在版编目（CIP）数据

表面的深度：中法联合教学教案 绘画·空间·设计/法国巴黎玛拉盖国立高等建筑学院，法国诺曼底国立高等建筑学院，中国东南大学建筑学院编著．—南京：东南大学出版社，2011.9

（高等院校建筑·艺术设计专业教学改革丛书）

ISBN 978-7-5641-3010-7

Ⅰ．①表… Ⅱ．①法… ②法… ③中… Ⅲ．①建筑设计-教学研究-高等学校 Ⅳ．①TU2

中国版本图书馆CIP数据核字（2011）第187361号

表面的深度	(法) Philippe Guérin 等 著
The Depth of	隽石(中国)文化传播机构
the Surface	JUNSTONE 策划

出版策划　　宋　磊　　王　盈
责任编辑　　宋华莉
视觉设计　　吴雪颖
责任校对　　徐　潇
出 版 人　　江建中

东南大学出版社出版发行（南京市四牌楼2号　邮编210096）
全国各地新华书店经销　　江苏凤凰扬州鑫华印刷有限公司印刷
开本：720mm×960mm 1/16
印张：12
字数：248千字
版次：2011年9月第1版　2011年9月第1次印刷
书号：ISBN 978-7-5641-3010-7
定价：60.00元

编委会

策 划：宋　磊　Song Lei
　　　　王　盈　Wang Ying
　　　　吴雪颖　Wu Xueying
　　　　东南大学建筑学院中法联合教学全体教师

编　委：（法）Philippe Guérin
　　　　龚　恺　Gong Kai（东南大学建筑学院副院长、教授，Professor, and Vice President of the School of Architecture, Southeast University）
　　　　曾　琼　Zeng Qiong（东南大学建筑学院环境艺术设计系副主任、副教授，Vice Professor, and Vice Director of the Envionmental Arts Design Department of the School of Architecture, Southeast University）
　　　　赵　军　Zhao Jun（东南大学建筑学院环境艺术设计系教授，Professor, Environmental Arts Design Department of the School of Architecture, Southeast University）
　　　　方晓珊　Fang Xiaoshan（东南大学建筑学院环境艺术设计系副教授，Vice Professor, Environmental Arts Design Department of the School of Architecture, Southeast University）
　　　　沈　颖　Shen Ying（东南大学建筑学院环境艺术设计系讲师，Lecturer, Environmental Arts Design Department of the School of Architecture, Southeast University）
　　　　朱　丹　Zhu Dan（东南大学建筑学院环境艺术设计系讲师，Lecturer, Environmental Arts Design Department of the School of Architecture, Southeast University）
　　　　张　蕾　Zhang Lei（东南大学建筑学院环境艺术设计系讲师，Lecturer, Environmental Arts Design Department of the School of Architecture, Southeast University）
　　　　戴　斐　Dai Fei（东南大学建筑学院环境艺术设计系讲师，Lecturer, Environmental Arts Design Department of the School of Architecture, Southeast University）
　　　　胡碧琳　Hu Bilin（东南大学土木学院讲师，Lecturer, Civil Engineering College, Southeast University）

作者简介

菲利普·葛汉（Philippe Guérin），1952年生于法国勃艮第省。

由于其所受的建筑学教育、作为画家的人生历程以及当前在现场艺术领域的经历，作为艺术家和教育者的实践，他逐步发展出一系列理论性及实践性的课题，这些课题以一种交叉的方式，试图在历史、现代与当代这三者之间建立一种本质联系。

自幼年起，他在艺术学校学习绘画，并钟情于博物馆和书籍里的绘画。17岁时，他受到了蒙德里安作品的影响，并于次年开始在巴黎学习建筑。1976年，他以"艺术与建筑关系之探索"为毕业论文，获得建筑师文凭，之后，他选择投身于绘画，并经常在欧洲展出作品。1995年起，他在法国的建筑学校任教，并于近期开始在意大利以及中国授课。自2002年以来，他举办了各类相关主题的讲座。

人性之存在，始终是他关注的核心。

教学：

巴黎瓦勒·德·塞纳国立高等建筑学院，巴黎，2011年至今

诺曼底国立高等建筑学院，鲁昂，2008—2011年

UNICA，建筑学院，Cagliari，意大利，客座教授，2010—2011年

巴黎玛拉盖国立高等建筑学院，巴黎，2000—2008年

巴黎威勒敏建筑学院，巴黎，1995—2000年

巴黎市艺术及建筑高等职业学校，1995—1997年

Philippe Guérin was born in Burgundy (France) in 1952.

Considering his training as an architect, his career as a painter and his recent experience with live arts, he has gradually been developing through his commitment as an artist and teacher some theoretical and practical inquiries which attempt to define essential connections between historical, modern and contemporary aspects. He began to study drawing at an early age, with a passion for paintings he saw in museums and books.

At 17 he discovered the work of Piet Mondrian and, the following year, he started to study architecture in Paris. Following his graduation in 1976 specializing in the links between art and architecture, he chose to commit exclusively to painting and frequently exhibited his works throughout Europe. In 1995, he set up a course for architecture schools in France, and recently, he implemented the same model in China and Italy. He has been lecturing on these subjects since 2002.

Human presence is always present at the core of his concerns.

Teaching Experience:

Architectural National Superior School Paris-Val de Seine, Paris, 2011 – present

Architectural National Superior School of Normandy, Rouen, 2008 – 2011

UNICA, Faculty of Architecture, Cagliari, Italy, Invited professor, 2010 – 2011

Architectural National Superior School Paris-Malaquais, Paris, 2000 – 2008

Architectural School Paris-Villemin, Paris, 1995 – 2000

Art & Architecture Professional Superior School of Ville de Paris, 1995 – 1997

Philippe Guérin est né en Bourgogne (France) en 1952.

Au regard de sa formation d'architecte, de son parcours de peintre et de ses expériences actuelles avec les arts vivants, il développe progressivement par un engagement d'artiste et d'enseignant, des questionnements théoriques et pratiques, qui, de manière croisée, tentent de prendre en charge des rapprochements essentiels, ou qui peuvent le devenir, entre apports historiques, modernes et contemporains.

Dès son plus jeune âge, il étudie le dessin dans une école d'art, se passionnant pour la peinture dans les musées et à travers les livres. A 17 ans, il découvre l'œuvre de Piet Mondrian et, l'année suivante, commence des études d'architecture à Paris. Après l'obtention du diplôme d'architecte en 1976, dans lequel il explore certains rapports entre art et architecture, il choisit de se consacrer entièrement à la peinture et expose régulièrement en Europe. A partir de 1995, il installe un enseignement en école d'architecture en France et, plus récemment, en Italie et en Chine. Depuis 2002, il donne des conférences sur ces thèmes.

La présence humaine revient toujours au cœur de ses préoccupations.

Enseignement:

Ecole Nationale Supérieure d'Architecture Paris-Val de Seine, Paris, 2011 – présent

Ecole Nationale Supérieure d'Architecture de Normandie, Rouen, 2008 – 2011

UNICA, Facoltà di Architettura, Cagliari, Italie, Professeur Invité, 2010 – 2011

Ecole Nationale Supérieure d'Architecture Paris-Malaquais, Paris, 2000 – 2008

Ecole d'Architecture Paris-Villemin, Paris, 1995 – 2000

Ecole Professionnelle Supérieure d'Art et d'Architecture, Ville de Paris, 1995 – 1997

序 言

记建筑学院的一次美术联合教学

每年秋季开学初的短学期，原是建筑学院美术教学的实习期。这时节，大二的学生会开心地背上画夹，跟随教师去一些风景秀丽之处写生作画。2010年，情况有所不同，学院邀请了法国巴黎玛拉盖国立高等建筑学院、诺曼底国立高等建筑学校的葛汉教授来校作了一次美术联合教学。

这里面有一个大背景，东南大学建筑学院近十年来，在本科教学中坚持"国际化"的方向，在建筑设计、城市设计和建筑技术等方面每年都要进行数次的联合教学，而基础教学中一个重要的方面——美术教学，以前却从未进行过，因此，美术学的教师很有意愿进行类似的尝试，在赵军教授等人的推动下，终于形成了此次的联合教学。

在我和葛汉教授的交谈里得知，他在大学里学的是建筑设计，毕业后却从喜好出发选择了绘画，因此就具有了建筑师和画家的双重身份和双重学术背景。学设计的人如何学美术，应该如何教，我觉得他这样背景的人是有深刻体会的。

整个联合教学我作为观察员大概参加了四次。第一次是葛汉教授刚到学校，作为礼节性的接待在赵军教授等一干教师的陪同下我们见了面。初次见面我就很是诧异：他的英语居然说得如此流利（因为在我印象中，法国人一般是不愿意说英语的）。想到他和我们的学生在接下来的一段时间交流应该不成问题了，他也大致谈了他的教学计划和方法——讲课和改图穿插进行；第二次旁听了他的第一节讲课，内容既有美术的历史，也有新的作品，可能是为了表达清楚或是让学生完全明白，他上课还是用上了翻译，幻灯片很精美，看得出是做好了准备；第三次是在教室改图，我看到有不少学生围在他身边，也就没凑近去打扰他们的讨论；最后一次是参加教学的最终展览，这是最热闹的一次，像过节一般，所有的教师和同学都来到布置好作业的评图室，听葛汉教授的逐一讲评，这本书主要展示的也就是这次的成果。

在我的观察中，葛汉教授的教学方式和我们以往的教学有很大的不同。我们的美术教师大多毕业于美术院校，教的过程中不自觉地也就延续了美术院校的教学方法，纸面的作业较多，强调基本功训练的较多，而葛汉教授在教学中，设计的成分更多，他会鼓励学生用各种各样的材料去进行创作。另外，讲课也是他教学的重要组成部分，在讲的过程中，有些观念和方法不知不觉地渗透到学生的作业中。

课程结束后我随机调查了下，绝大多数的教师和学生都觉得他教得非常有趣，和他们平时做的不一样，也带来了思索和想象的空间。我觉得，这就是联合教学的魅力，不同文化的交流、碰撞会给我们的教学带来新的内容，不同的教学方式会给我们的教师以新的启发。

最后，感谢王建国院长对此次活动的大力支持，感谢葛汉教授、王盈先生有意义的教学，感谢隽石（中国）文化传播机构宋磊先生的大力支持，感谢赵军教授、曾琼副教授、方晓珊副教授、沈颖、朱丹、张蕾、戴斐、胡碧琳等美术学教师的组织和辅导，以及全体参加的同学，因为有了大家的积极参与，这次活动的成果才能集结成书呈现给读者。

东南大学建筑学院副院长、教授 龚恺
2011.3.18

Preface

Looking Back on the School of Architecture's Experiment in Joint Teaching

Every year at the beginning of the fall semester, the School of Architecture holds its Arts Education Internship. During this period, second-year students throw their easels on their backs and follow the instructor to a place of magnificent scenery and begin to capture the world in their art. 2010 was a little different than usual, as we invited professor Philippe Guerin from the Ecole Nationale Supérieure d'Architecture de Malaquais and Ecole Nationale Supérieure d'Architecture de Normandie to cooperate with us in the joint teaching of art.

There is a large context to this special change. For the last 10 years, Southeast University has persisted in "internationalizing" the curriculum directions with architectural design, urban design, architectural techniques, and other joint-taught classes, multiple times per year. However, such a core curriculum as arts education had never before been attempted. Arts instructors have been hoping to execute a similar experiment, and thanks to professor Zhao Jun's propelling, last year's joint teaching program of art was initiated last year.

From my discussions with Professor Guerin, I came to learn that after graduating from his architectural design studies, this Frenchman followed his long-time hobby and chose to continue his studies in the field of visual arts. Accordingly, with the double title of architect and artist, his background in both subjects prepares him well to understand how designers learn art, and therefore how they should be taught.

Over the entire teaching period, my role was that of a casual observer, participating 4 times in total. When Professor Guerin first arrived at the school, as a part of the formal welcome, accompanied by Professor Zhao Jun and the rest of the professors, we met for the first time. At once, I was astounded at the fluency of his English (I was under the impression that French people aren't fond of speaking English), and realized that he would have no problem at all communicating with our students. He also outlined his teaching plan and method of alternately lecturing and making suggestions to improve students' work. The second time was when I sat in on his first class. His lecture ranged from art history to pieces of contemporary art, and, whether it was to make sure his point was delivered clearly, or that students could completely understand. He lectured with a translator. His slides were exquisite and it was apparent that he had done a lot of preparation. The third time I showed up in the classroom he was making suggestions on the students' work, but as there were a large group of students crowded around him I thought it was best not to butt into their discussion. The last and most exciting time was participating in the final exhibition of the class's work. With excitement reminiscent of the Spring Festival, all teachers and students crowded into the display room covered with students' work and listened to Professor Guerin give his comments one by one. The purpose of this book is mainly to display the fruits of this exhibition.

According to my observation, I found that Professor Guerin's instructional methods had many points different from our usual way of teaching. Most of our art teachers graduated from institutes of art, and whether they realize it or not, this comes out in their way of teaching: a large amount of paperwork emphasizing training in the fundamentals. Professor Guerin's method, on the other hand, contained more elements of design, encouraging students to use a wide variety of materials to complete their creations. Also, lectureing was an important element of his class, while during the lectures, some ideas and methods unconsciously had influence on the students' artwork.

After the course, I took the opportunity to take a survey and found that the vast majority of both students and teachers felt that Professor Guerin's teaching was quite interesting, different from their usual way of doing things, and brought many opportunities for imagination and speculation. In my mind, this is the virtue of joint teaching: the coming together and communication between different cultures bring new horizons to our educational content, and exposure to a new educational method gives our instructors new inspiration.

Finally, I'd like to thank Wang Jianguo, President of the school, for his enthusiastic support of this project, and Professor Guerin and Wang Ying for their meaningful teaching. Thanks to Mr. Song Lei from the Junstone Cultural Communication Co.Ltd. for his tremendous support. Thanks to Professor Zhao Jun, Associate Professors Zeng Qiong and Fang Xiaoshan, as well as Shen Ying, Zhu Dan, Zhang Lei, Dai Fei, Hu Bilin, and the rest of the arts instructors for their help in organization and guidance. Most importantly, thanks to all the students who took part in the activity, since it was their active participation that made the fruit of this activity able to be published in this form for the enjoyment of its readers!

Gong Kai (Professor, and Vice President of the School of Architecture, SEU)   March 18, 2011

前 言

参与这个介绍欧洲文化（到中国）的项目是一种亲缘的行为和信仰。这种欧洲文化是大家所希望接触的，也是我们（作为欧洲人）至今还略微自豪的（其具有一些成就）。中国教师和年轻学生，今天还需要（这样的外来文化），因为也许将来还可以用它来实现大家的理想，并将此尽可能"内化为文化"，以便能和你们自己的、曾经有所断裂的文化本源重新建立起联系，因为你们自己的文化本源似乎没有被很好地传承下来。此外，这种"移植"可能对我们自己来说也是有益的，尤其是因为我们拥有的文化（知识）以及更多的真诚，并有着将之实现的决心。历史上各种（文化的）"杂交"总是不断地出现着。不管怎么说，我感到这可能非常重要，而且在我看来，我们可以用相同的语言进行交流。很显然，这会让我们的对话变得意趣盎然，也使这种对话变得更有可能。因为要知道，中国和欧洲这两者并不总是对各自的情况能互相作出相同的分析（文化的差异）。但正是这种认识的多样性，可以让我们在相互的交往中做到更加精确，更加中肯。为什么不这样做呢？！这样能让我们对大家的未来了解得更加透彻。

菲利普·葛汉

Introduction

Participating in this European contribution, to which you invited us and in which we still believe, represents a gesture of accomplishment and inner conviction which you, Chinese teachers and young students, will be able to put to good use in the future as a "breeding ground" in order to recreate links with your own culture – a culture that has unfortunately been mistreated at the end of the nineteenth and during the twentieth century. This "transplantation" will perhaps also be beneficial to us, all the more so as we will strive to achieve it with a little bit of knowledge and perfect honesty. The hybridisation of cultures has always prevailed throughout history. In my opinion this fusion was the most important part of our project, and I firmly believed that we would manage to speak a common language. That common language is what allowed for dialogue and for our exchanges to be so interesting, despite the fact that our points of view on our own situations in Europe and in China sometimes diverge. On the contrary, these differences of opinions can bring additional precision and pertinence to our present exchanges, and maybe even help have a clearer view of the issues that will drive us in the future.

Introduction

Participer de cet apport européen pour lequel vous nous invitez et auquel nous croyons encore un peu, est un geste d'accomplissement et de conviction intime dont peut-être effectivement, vous, enseignants et jeunes étudiants chinois, avez besoin aujourd'hui pour l'utiliser demain à votre fin en le transformant comme une "mise en culture" possible, afin de retisser des liens avec votre propre origine, qui fut malheureusement malmenée à la fin du XIXème et pendant le XXème siècle. D'ailleurs, cette "transplantation" sera peut-être bénéfique pour nous mêmes, et ce d'autant plus que nous aurons le cœur pour le faire avec un peu de savoir et beaucoup d'honnêteté. Les hybridations ont toujours été présentes dans l'Histoire. C'est en tout cas l'enjeu que je ressens comme potentiellement important et il m'a semblé que nous pouvions parler un langage commun. C'est bien sûr ce qui rend nos conversations très intéressantes et le dialogue possible en sachant que nous n'avons peut-être pas toujours exactement la même analyse sur nos propres situations réciproquement en Chine et en Europe. Mais justement, cette diversité des points de vue peut aussi nous conduire à de plus grandes précisions et pertinences dans les échanges et pourquoi pas, à plus de lucidité sur les enjeux que (qui) nous porterons pour le futur.

Philippe Guérin

目 录

- 000　**1　法国建筑院校中的艺术教育**
- 000　1.1　艺术与建筑
- 006　1.2　总体方法论和个人方法
- 012　1.3　我在建筑院校教授"表现艺术和方法"课的目标
- 014　1.4　在法国两所院校中的独特经验

- 038　**2　东南大学建筑学院教学活动**
- 038　2.1　教学活动的准备
- 038　2.2　日程安排及进展
- 046　2.3　定制课程的内容
- 054　2.4　学生们的反应和与他们的对话
- 056　2.5　与中方教师的对话以及作出可能的调整
- 057　2.6　法文原文

- 067　**3　作品分析**

- 130　**4　课后文集**
- 130　视觉设计基础教学　曾琼
- 134　中法联合教学的启示　赵军
- 140　超透视表达　沈颖
- 150　心智与图式　朱丹
- 158　横看成岭侧成峰　胡碧琳
- 164　关于加强美术教育在设计课程中功能的思考　戴斐

- 173　**附录　视觉设计基础教学大纲**

- 176　**后　记**

Contents

001	**1 The teaching of art in French architecture schools**
001	1.1 Art and architecture
007	1.2 General methodology and personal approach
013	1.3 My goals as an ATR teacher in architecture schools
015	1.4 Particular experiences in two French schools
039	**2 Workshop at the School of Architecture of SEU**
039	2.1 Preparation of the Nanjing's workshop
039	2.2 Day by day diary of events
047	2.3 Contents of the lectures
055	2.4 Students' reactions and dialogues
056	2.5 The dialogue with the Chinese teachers and eventual adjustments
057	2.6 Articles in French
067	**3 Analysis of work**
131	**4 Texts and other comments**
131	Fundamentals of Visual Design Zeng Qiong
135	Lessons Learned from Our Sino-French Joint Teaching Zhao Jun
141	A Perspective Expression Shen Ying
151	Psyche and Schema Zhu Dan
159	Different Perspectives Lead to Different Results Hu Bilin
165	On Strengthening the Role of Arts Education in Curriculum Design Dai Fei
173	**Appendix Syllabus**
176	**Epilogue**

1 法国建筑院校中的艺术教育

1.1 艺术与建筑

1.1.1 欧洲历史上的几个里程碑

在西方文化历史上，15世纪初的意大利，更确切地说是在佛罗伦萨，一场被后人称为文艺复兴的运动（布克哈特（Burckhardt），《意大利文艺复兴时期的文化》，1860）会集了很多的建筑师、雕刻家和画家，他们在之后便掀起了艺术表现准则的革命。这一新思潮的基础就是理论创新，即基于绘画手法的透视法——艺术设计，这一新概念之于艺术表现的作用犹如动词的用法之于雄辩术和诗歌。

我们所熟悉的"布扎"（美术）这一术语，19世纪才出现。历史上还存在着一种建筑风格叫"布扎"，即新古典主义，受到了法国的影响，后来走向全世界，包括美国，比如纽约和芝加哥这样的城市。中国南京东南大学的大礼堂（图1），就是符合了这一时期形式上的某些标准。

与之相反的是，同一时期的欧洲，尤其是在1905—1930年之间，现代主义的革命成果尤其丰硕。这种对于承袭自文艺复兴时期的传统表现手法及法则的重新跨越，促使艺术家们进行更多的试验和探索。粘贴、拼贴纸的使用，用于艺术展览的现成物品的发明，抽象艺术的选择，体现的是人类质疑世界的新自由……而且通过各种各样的方式，试图渐渐地从传统的绘画平面和传统的画框中走出来。

图1 东南大学大礼堂

后来被我们公开称为的"造型艺术"，促进了画家、雕塑家和建筑师之间开展深度的合作。包豪斯（Bauhaus）、风格派（De Stijl）以及构成主义就是对此的绝佳范例，并深深表达了其精髓。在我们撰写此文的时候，蓬皮杜艺术中心（图2）正在举办一场别开生面的展览，展出的正是皮埃·蒙德里安（Piet Mondrian）（图3）及风格派运动的作品。

其次，在二战期间，欧洲很多著名艺术家移居北美，尤其是纽约。因此，20世纪后叶的艺术从根本上来说受到了美国的影响。然而，北美和老欧洲之间的交流仍然非常密切。1977年，蓬皮杜中心就是以一场具有象征意义的题为"巴黎-纽约"的展览揭幕的。2000年9月，惠特尼博物馆（Whitney Museum）（图4）也正是以一个同样具有启示意义的标题，"美国世纪——艺术和文化1950—2000"，通过对美国艺术的重要回顾，来告别这个千年。

如今，到了2011年，我们显然必

1 The teaching of art in French architecture schools

1.1 Art and architecture

1.1.1 A few European landmarks

Historically, in Western culture, the artistic movement which started in Italy, and namely in Florence, at the beginning of the fifteenth century, and which was later to be called the Renaissance (Burckhardt, 1860, *The Civilization of the Renaissance in Italy*), has brought together architects, sculptors and painters who revolutionised the codes of artistic representation. The basis of this new movement was a theoretical innovation, namely perspective, rooted in the use of drawing – *Arti del Designo*. This new concept was to artistic representation what rhetoric and poetry was to the use of language.

The familiar terminology "Fine Arts" emerged much later in the nineteenth century. There was in fact a neoclassical and French-influenced architectural style called "Fine Arts", which was exported all around the world, including in the USA, such as NYC or Chicago. The premises of the Southeast University (Figure 1) include quite a few formal criteria from this period.

图2 法国巴黎蓬皮杜艺术中心

Paradoxically, in Europe, during the same period between 1905 and 1930, the upheaval of Modernity was very prolific. This emancipation from the traditional codes of artistic representation, inherited from the Renaissance, prompted artists to make numerous experiments and discoveries. The use of pasted paper, the invention of the ready-made, the use of abstraction, were all new ways through which men challenged the world around them. Through various means they all contributed to progressively bringing the painting out of its traditional frame.

What were later known as the "fine arts" played an important role in bringing painters, sculptors and architects together. Le Bauhaus, De Stijl or Constructivism are all noteworthy examples of this collaboration. As we are writing this text, an outstanding exhibition is taking place at the Centre Georges Pompidou(Figure 2), presenting the works of Piet

图3 皮埃·蒙德里安（Piet Mondrian）自画像

须用这种充满全球化的眼光来关注正在发生的变化，关注未来的转变，而中国也将在这转变过程中起到举足轻重的作用。

1.1.2 法国体制上的几个里程碑

1870年，法国在第三共和国时期，有了公共教育、宗教和艺术部。1881年，儒尔·费利（Jules Ferry）〔图5〕出任了公共教育、宗教和艺术部的新部长，而且，实际上也正是在19世纪，艺术作品的概念获得了极大的发展，并且不再优先反映宗教灵感。

随后，到了1959年，安德烈·马尔罗（André Malraux）〔图6〕首次被戴高乐任命为国务大臣，主管文化事务，并于1962年被再次委任，从而为文化赋予了一种新的自主。

1981年，法国的这种特点，为弗朗索瓦·密特朗（François Mitterand）所再次强化，尤其是在杰克·朗（Jack Lang）第一次被任命为文化部部长的时候，这种影响力更是得到了加强。

如今，由于这个历史背景，我们成立了两个截然不同的部委延续了这样的状况：1932年成立的国民教育部，以及文化和交流部，后者自1997年起往往采用这一名称。

法国艺术和文化机构设置的变化

图5 儒尔·费利（Jules Ferry）

意义重大。然而，你仍可以在巴黎波拿马街的历史街区里发现巴黎国立高等美术学院（ENSBA）的校址〔图7，图8〕。这所学院成立于1819年，前身为皇家艺术学院，随后在1883年改为帝国学院，直至1968年，该院一直都在进行建筑学以及绘画、雕塑和雕刻等专业的教育。

实际上，1968年爆发的"五月风暴"标志着法国在政治和认识论上的决裂，这一事件具有历史意义。这种决裂将学科体系创新引入了巴黎国立

图4 美国惠特尼博物馆

Mondrian(Figure 3) and De Stijl movement.

Then, during the Second World War, many great European artists emigrated to North America, and mainly to NYC. On that account, the art of the second half of the twentieth century showed evidence of a strong American influence. Nevertheless, exchanges between North America and the old continent remained very active. In 1977, the Georges Pompidou Centre was inaugurated with an emblematic exhibition named *Paris-New York*. It was with an equally significant title for its retrospective on American art – *The American Century - Art & Culture 1950 - 2000* – that the Whitney Museum (Figure 4) ended the millennium.

Today, in 2011, we need to take globalisation into account when we observe the ongoing changes, and the future transformations, in which China will certainly play a very important part.

1.1.2 Some French institutional landmarks

In 1870, during the Third Republic, France had a Ministry for Public Instruction, Cults and Fine-arts. In 1881, Jules Ferry(Figure 5) was appointed Minister for Public Instruction, Cults and Fine-Arts. It was indeed during the nineteenth century that the notion of work of art evolved tremendously to cease to be mainly an answer to religious aspirations.

Later on, in 1959, and then again in 1962, André Malraux(Figure 6) was appointed by the Général de Gaulle as Minister of State, in charge of Cultural Affairs, thus giving culture a new form of autonomy.

This French distinctiveness was enhanced in 1981 by François Mitterrand, through a considerable budget, but also through the appointment of Jack Lang as Minister of Culture.

Nowadays, owing to this historical background, we are endowed with two separate ministries, the Ministry of National Education, introduced in 1932, and the Ministry of Culture and Communication, which has existed under this name since 1997.

The evolution of French institutions of art and culture has been quite significant, but one can still find the National Superior School of Fine-arts(Figure 7,8) in its historical site, in rue Bonaparte, in Paris.

图7 巴黎国立高等美术学院内部

图6 安德烈·马尔罗 (André Malraux)

图8 巴黎国立高等美术学院（ENSBA）校址

图9 巴黎玛拉盖国立高等建筑学院

高等美术学院内部，随后，建筑教育由8个互相之间差异很大，且分布于法国各地的建筑教学单位（UPA）来进行。此外，安德烈·马尔罗首先做出了这个决定，拒绝学院派和当时非常强势的"工艺美术"风格。几年当中，建筑教育更多依仗的是装备部，直至20世纪才归到文化部。

总体说来，这种组织架构如今继续存在。然而，教学单位在此时已成为了国立院校，后来便成为了国立高等建筑院校。

这些院校中的最新成员——巴黎玛拉盖（Paris-Malaquais）国立高等建筑学院，由一群教师于2001年，在这个历史古迹中成立。他们的目标是试图重新构筑和巴黎国立高等美术学院（ENSBA）的合作。尽管如此，两者的分离却依然非常明显。

1.1.3 表现的艺术和方法

1968年后，全法国在每个院校内部，我们从所谓的"垂直"工作室教学转向了"水平"的、跨专业的、围绕着设计的主题教学。总共有6个教学领域，并通过全国性的考试选择聘用授课教师：

· 建筑设计的理论与实践；
· 建筑史和建筑文化；
· 建筑科学和技术；
· 城市和区域；
· 表现的艺术和方法；
· 人文和社会科学。

根据法国建筑学院内的正式名称——表现的艺术和方法（ATR），它本身又可以分为两个子领域：

· 表现方法；
· 造型和视觉艺术。

这种分类有时会产生一定的危害，因为这往往会导致工具和思想之间的差距过大。因此，有时会使得学

Founded in 1819 as the Royal school of fine-arts, then renamed to the Imperial school in 1883, it used to be home to the teaching of architecture in conjunction with the teaching of painting, sculpting and carving until 1968.

Indeed, the events of May 1968 represented a major political and epistemological fracture in France. This fracture caused the breaking up and the scattering of ENSBA fields of study. Afterwards, the teaching of architecture was provided on different sites by 8 distinct Architecture Pedagogical Units (UPA in France) throughout France. André Malraux initiated these changes, in response to the violent protests against academicism and the "Fine arts style". For a few years the teaching of architecture depended on the Ministry of Equipment, but quickly returned under custody of the Ministry of Culture.

To a large extent, this organisation is still in place today. Nevertheless, the pedagogical units have become national schools, and then national superior schools of architecture.

One of the latest of these new schools is the National Superior School of Architecture Paris-Malaquais(Figure 9), created in 2001 on the historical site by a group of teachers. They wanted to restore the possibility of collaboration with the National Superior School of Fine-Arts. Despite these efforts, the separation between them remains very strong.

1.1.3 The field of ATR

Within each school, after 1968, we have moved from teaching in so-called "vertical" workshops, to so-called "horizontal" interdisciplinary courses gathered around a common project. There is a total of six different disciplinary fields which are subject to specific recruitments through national examination:

· Theories and processes of architectural conception
· Architectural history and culture
· Science and techniques linked to architecture
· Cities and territories
· Arts and techniques of representation
· Human and social sciences

According to the official classification within the schools of architecture in France, the ATR (Art and Techniques of Representation) field is also divided into two sub-fields:

· Techniques of representation
· Plastic and visual arts

This division can sometimes be a bit harmful, as it can lead to the existence of too great a gap between tools and ideas. Because of that, students are sometimes unable to express their thoughts, or on the contrary, be excessively reliant on techniques which impose premature constraints on their projects.

In any case, the assimilation of theoretical knowledge is generally improved when it works hand in hand with some practical and personal experiences. Students can then more easily pay attention to the skills which will prove to be the most useful and interesting for their studies, and progressively position themselves towards the professions of architecture in accordance with conceptual as well as technical choices. Concerning this

生常常不能表达他们的思想，或是与之相反，太依赖于对他们的设计过程产生影响的技术。

不管怎么说，当理论知识与实践相结合时，一般都比较好吸收。因此，学生们可以更容易地注意到对他们的研究有用的技巧，并且在建筑设计这一行业中，根据不同的选择（这种选择既是观念上的也是技术上的），逐步地定位自我。此外，在面对建筑行业以及城市演变的时候，同样需要让他们能够更好地拓展将来的能力，从这个意义上来讲，学生有着实质的需求。

1.2 总体方法论和个人方法

1.2.1 画家和建筑师：双重教育

从11岁至17岁，我在一个美术学校学习绘画。通过书本和博物馆（图10），我对画画产生了极大的热情。在17岁的时候，我接触到了蒙德里安（Mondrian）的作品。次年（也就是1970年），我开始在巴黎学习建筑。这给了我观察艺术和建筑之间某些关系的好机会，而且直至今天，我仍然通过自己的职业活动继续着这种探索（图11）。

1976年，当我建筑师文凭论文答辩（论文标题：建筑与梦想——绘画和建筑之间的关系）的时候，我邀请安德烈·费明捷*（André Fermigier）当我的评委会成员。他接受了我的请求，但却提醒我说："绘画和建筑是假朋友"，并还说："你以后会发现，这并不容易。"事实确实如此，很多年之后，我才能不用一个建筑师的眼光，而是像画家一样去看待绘画空间。这并不是同一种空间。而且从本质上说来，这也不是同一个学科。

*安德烈·费明捷，1923—1988，历史学家、艺术评论家、教授。他曾为很多著名画家撰写过多部著作，如毕加索（Picasso）、波纳尔（Bonnard）、库尔贝（Courbet）等。从20世纪60年代至80年代初，他还撰写建筑及城市规划的专栏文章，最后汇集成书，名为《巴黎的战争》（图12）（伽利玛出版社，1992）。

图10 法国巴黎罗浮宫

图11 作者工作室

matter, student requests are actually quite substantial and they have to be answered satisfactorily so as to allow the students to develop the most efficient abilities necessary to face the evolution of architecture professions and of the city itself.

1.2 General methodology and personal approach

1.2.1 Painter and architect, a double training

Between the ages of 11 and 17 I studied drawing in an art school, and through books and museums(Figure 10) I developed a passion for painting. At 17 years old I discovered Mondrian's work. The year after that, in 1970, I started my architecture studies in Paris. This allowed me to observe some links between art and architecture and today, through my professional activities, I am still engaged in this research(Figure 11).

In 1976, before doing my academic defense *(title: Architecture and the dream,* subtitle: *The relations between painting and architecture)*, I asked André Fermigier* to be part of my committee. He accepted but warned me: "painting and architecture are faux amis" and added "you'll see, it won't be easy". Indeed, it took me years to star using the painting space as a painter rather than as an architect. It is not the same space. It is definitely not the same job.

*André Fermigier (1923-1988) was a historian, art critic and professor. He was the author of several works about important painters—Picasso, Bonnard, Courbet etc. From the 1960s to the beginning of the 1980s he wrote columns about architecture and urbanism, which were later collected in a book (*The Battle of Paris*, Gallimard, 1992, Figure 12).

From 1976 to 1994 I decided to devote myself strictly to painting, and exhibit my works in Europe. Nevertheless, and this is quite an important point, all my exhibits showed the relation between the painting and the architectural space. It was only in 1995 that I decided to teach in different architecture schools. Since 2002, through lectures and workshops, my artistic and pedagogical approach become more precise as I started mixing increasingly my professional activities.

As mentioned here-above, Modernity, mainly during the interwar period, has certainly renewed several obvious links between art and architecture, but today we need to consider the dcbate from a 21st century perspective, and discuss the evolution of what we call nowadays

图12 《巴黎的战争》

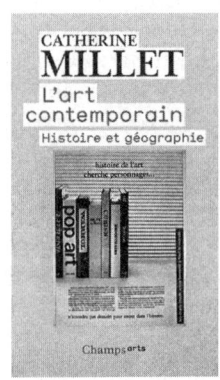

图13 《当代艺术、历史及地理》

从1976年至1994年,我决定把所有的时间用来画画,然后在欧洲展出。然而,很重要的一点是,我的所有这些展览都将绘画和建筑空间联系了起来。仅从1995年起,我才决定在几个建筑院校里从事教学工作。2002年以来,通过各种讲座和研讨会,我越来越多地将专业活动交叉起来,使得艺术方法和教学方式日趋成熟。

现代性,我们之前已经提到过,尤其是在两次世界大战期间,某种程度上来讲,无疑地重新搭建了艺术与建筑之间交流的桥梁。而如今,到了21世纪,我们需重新考虑艺术和建筑之间的争论,考虑我们今天所说的"当代艺术"。此外,我们还能非常理性地去怀疑和检验20世纪80年代以来一直占据国际舞台的"当代艺术"这个说法的历史定义及其永恒的含义。

为了具体说明这些术语,并丰富我们的讨论,我们引用评论家(《Art-Press》杂志)、作家卡特琳娜·米莱*(Catherine Millet)的一句话:"因为并不是艺术在设计、资料图像、服饰等等中以功用性为目标,而是这些活动试图获得艺术的身份。同样,我们应坚持这个定义:当代艺术是个引力极。不多也不少。"

*卡特琳娜·米莱(Catherine Millet). 当代艺术、历史及地理[M]. 巴黎:弗拉马里翁出版社(Flammarion),2006. (图13)

最后,我们还注意到,2010年11月于里昂举办的一次研讨会——"艺术和建筑"上,玛蒂娜·布施*(Martine Bouchier)作为建筑师、美学博士、巴黎瓦勒·德·塞纳建筑学院教授,提出"建筑师和艺术家并没有合作的使命"。当然,除了对这种横向性作出有益挑战之外,玛蒂娜·布施准确而公正地指出了这种依然存在的混淆,并且根据其在体制内的互相关系,以及在语义上将艺术作品过度改称为文化物品这种方式,来明确区分了这两个截然不同的概念。

*玛蒂娜·布施(Martine Bouchier). 艺术并非建筑[M]. 巴黎:Archibooks出版社,2006. (图14)

如今,从文化物品到众所崇拜的物品,两者之间的距离变得越来越近。

图14 《艺术并非建筑》

"contemporary art". We can question and examine the historical definition and the persistence of the phrase "contemporary art" — a phrase which has prevailed on the international stage since the 1980s.

For the sake of clarity and to ensure the quality of our debate, we will refer to a quote by Catherine Millet*, art critic (*Artpress*) and writer : "...because art is not aiming at functionality in the fields of design, journalistic photography, fashion, etc., these disciplines on the contrary are trying to achieve the status of art. Therefore, we can state that contemporary art is a pole of attraction. Nothing more, nothing less."
*Catherine Millet. *Contemporary Art, History and Geography*. Flammarion, 2006. (Figure 13)

Before concluding, I would like to mention a statement by Martine Bouchier*, architect, HDR doctor in aesthetics and professor at the ENSAPVS, who declared during the "Art and Architecture" conference in Lyon in November 2010 that "the architect and the artist were not inclined to collaborate". A provocative and interesting statement, Martine Bouchier's comment also accurately describes the prevailing confusion. Furthermore, it points out two distinct notions by slipping semantically from the work of art to the cultural object according to their mutual links with the institutions.
*Martine Bouchier. *Art is not Architecture*. Archbooks, 2006. (Figure 14)

Nowadays, the distance between the cultural object and the object of cult is thinning out.

1.2.2 Transversality in teaching

Considering my training as an architect, my career as a painter, and my recent experiences in the practice of artistic activities, I am now able to progressively melt up together my artistic abilities and my pedagogical works by developing interdisciplinary theoretical inquiries about representation in the two-dimensional as well as in the three-dimensional space.

To sum up, my thoughts on the place allotted to art classes in architecture schools, can be divided into two main categories:
• autonomous artistic teaching given as specific classes and which has no direct implication in an architectural project, but which instead focuses on acquiring tools and knowledge regarding techniques of representation, and then of general culture, as the two are essential to the training of architects.

• Transdisciplinary teaching providing artistic training which will be included in the elaboration of an architectural project. These courses will not necessarily be operational straight away, but bring up the idea of a return to project work and learning through personal experience and knowledge which enrich the creative process.

Interdisciplinarity, and therefore transdisciplinarity, both imply the existence of disciplinary fields and the development of specific thought processes. For it to be truly beneficial to the student's research, the journey needs to be completed with a return to the initial discipline in order to go over its adjoining rules. It is a crucial point, because too many students (as I have witnessed in my classes), attracted by the changes brought by transdisciplinarty, can be tempted by the new paths and

1.2.2 教学间的横向性

我所受的建筑学教育、作为画家的经历以及如今从事的与现场艺术相关的各种经历,让我通过开发这种把涉及平面空间以及三维空间表现的理论问题交叉起来的方法,逐渐将我的艺术实践和我从事的教学工作融合起来。

我关于建筑院校各个艺术学科地位的观点,总结起来,可以将他们分为两个主要方向:

· 以特定课程进行的自学式艺术教育,这与具体的建筑设计方案没有直接关联,但首先会让学生获得与表现技法相关的工具和知识,然后让他们学到文化概况,这两者在建筑师的职业培训中是必不可少的。

· 跨学科教育,教授的是艺术的曲折之法,然后融入到建筑方案的设计过程中。这种教学并不一定立即就具有操作性,而是提出了这样的问题:回归到方案,消化所获得的经验和知识,从而丰富自己创造性的过程。

从本质上讲,跨学科性和超学科性意味着各种学科领域的存在,以及它们所特定的思想的发展过程。为了让这些对学生的研究能够真正有益,教学的历程应该最终全部回归至出发时的领域,让他们能再次研究相关的法则。这是非常重要的一点,因为太多的学生,往往为跨学科带来的变化所诱惑(这是我在自己的教学中发现的),他们有时受到了某些新的方法和行为的吸引,而这是他们在自己的个人探索中没有时间去理解和消化的。这不是说要侵犯在相邻领域中进行探索发现的自由,而是应当充分理解和更好地使用那些本身就存在于建筑学中的自由。

1.3 我在建筑院校教授"表现艺术和方法"课的目标

如今,传统的方法和创新的技术手段在艺术领域并驾齐驱。这种创新的技术,我们在日常生活领域中都能发现,它以跟我们所有人都熟悉的内容作为出发点——物种繁殖。

事实上,创造和繁殖之间的关系,带来了移植或部分克隆人体器官的多种可能性,从而回避了质疑源头、叛离以及重新表现等一系列的问题。

因此,出于这几点,我对学生们提出几个明确的目标:

(1) 理解欧洲属性

我的教学源自于表现艺术和方法的几个基本方面。它们深深地扎根于西方艺术史。并且通过我们电脑这个窗口,借助于平面屏幕,增加了一个面向世界的开放视野。

它们都归结到一个统一的题目下——表现、再现(描绘)、存在。

· 空间中的一点(绘画教研室,三年制学士)

· 从历史上的深度(表现)到当代的平面性(表现)(理论和实践教研室,三年制学士)

(2) 现场艺术方面的训练

我的教学贯穿了几个艺术领域。人体则成为了思想特有的载体。我们通过它来理解建筑的空间。这种体验被沃尔特·本雅明*(Walter Benjamin)称为"触觉接受"。这和视觉一样,是我们体验接受建筑的基石。

*沃尔特·本雅明(Walter Benjamin),《机械复制时代的艺术作品》,1939,(1936年第一版)。

· 建筑中的人体(主题教学和工作组模式教学,三年制学士和两年制硕士)

· 转化为作品(理论教学和指导习作,三年制学士和两年制硕士)

不可否认,创新的技术手段在今天的建筑院校中已是大势所趋。但也因为如此,在这样的教学环境中进行创新,也意味着能够掌握浩瀚的理论和实践知识。这样可以最终承担起在历史、现代和当代之间创造本质联系的任务。

我的教学工作正是按照这种"三部曲"展开的,并随着学校的不同而进行调整。通过它们,我试图解释某些显而易见的或是现实存在的矛盾。正如马塞尔·普鲁斯特*(Marcel Proust)说的:"明日的偏见",为的是帮助学生在面对当今挑战的时候能够进行自我定位,这首先是在建筑院校里,当然,接着就是在外面,面对着日新月异的学科职业的发展。

*马塞尔·普鲁斯特(Marcel Proust)(1871—1922),"今日的矛盾就是明日的偏见",《快乐与日子》,1896。

behaviours which they did not have the time to fully assimilate and comprehend. This does not imply that the freedom brought by artistic teaching has to disappear from the pedagogical journey, but rather we should fully understand and put to good use the freedom that already exists in architecture.

1.3 My goals as an ATR teacher in architecture schools

Nowadays, art calls for the use of traditional approaches alongside with the innovating technological approaches which we can find in the everyday aspects of life, starting with one we all are familiar with, the reproduction of species.

Indeed, the link between creation and procreation brings up various possibilities of transplant or partial duplication of the human body, and alters the debate about origin, transgression and representations.

These ideas are the starting point from which I developed a few clear objectives for the students:

(1) Understanding a European lineage

My teachings are derived from fundamentals of the field of ATR and are deeply rooted in the history of Western art. Furthermore, they contribute to complete the open view we have of the world through the flat screens of our computers.

They are grouped under a generic title: Presentation, representation, presence.

·A point in space (drawing workshop, bachelor's degree)
·From historical depth to the contemporary plan (theoretical and practical workshop, bachelor's degree)

(2) Working at human scale

My courses stretch across a number of artistic fields. The human body is the primary medium through which thought can apprehend the space of a building by experimenting what Walter Benjamin* calls "the tactile reception" which is as important as sight in establishing our reception of architecture.

*Walter Benjamin (1892-1940). *The Work of Art in the Age of Mechanical Reproduction*. 1939 (first published in 1936).

·The body to the building (thematic teaching and studio project, master's degree)
·Transformation at work (theoretical teachings and tutorials, bachelor's and master's degree)

It seems undeniable that the use of innovating technologies is inevitable nowadays in architecture schools. However, innovating in this pedagogical environment, also implies being able to establish a large enough keyboard of theoretical and practical knowledge so as to be able to eventually undertake the task of creating essential links between historical, modern and contemporary contributions.

My courses are based on this "trilogy" although they vary according to the different schools. Through them, I attempt to explain some apparent or real contradictions and perhaps some paradoxes, "the prejudices of tomorrow" according to Marcel Proust*, in order to prepare students to face the present issues—at first within architecture schools, and then in the outside world, with its evolution of the professions related to this constantly mutating field.

*Marcel Proust (1871-1922), "The paradoxes of today are the prejudices of tomorrow". *Pleasures and Days*, 1896.

1.4 在法国两所院校中的独特经验

1.4.1 法国诺曼底国立高等建筑学院于2008年实施的表现艺术和方法教学案例

《空间中的一点》

教学目的

该教学法是针对一年级的大学生设计的,为此,它具有决定意义,因为它为诺曼底国立高等建筑学院的学生奠定了艺术表现领域的基石。该教学法的第一堂课首先由教师讲授,在课上教师会提出西方空间的几个基本内容,此外,也正是以这样的课程,我们开始了中国南京东南大学2010年的教学活动。

该教学法的目的在于,保证学生获得初步的理论基础,以及根据不同的表现方式,尤其是根据绘画,获得实际操作经验,从而有利于:

· 实际尺度和艺术表现尺度之间的对比;

· 对三维向两维空间过渡以及二维向三维空间过渡形成更好的理解;

· 更好地获得表现艺术和方法的历史和理论知识;

· 获得与所采用的创作方法相适应的准确的观察和重构;

· 想象的自由度,其依赖于表现个人思想的工具;

· 对新领域、新学科、新艺术或新技巧进行探索,更有利于将人的身体作为对新空间进行研究的载体。

教学方法

在介绍的案例范围内,教师授课之后是参观鲁昂博物馆(图15),而且根据随意选择的一幅油画,要求学生们根据这幅油画,绘制几幅图画,随后为接下来的授课准备一个初步的立体模型。这个练习总体上持续5~6周,其中有三维的和二维的作业。随着学生们一步步展开自己的研究,教师给予他们的首先是集体指导,随后才是逐个指导。根据进度的不同,教师会鼓励每个学生逐渐找到一种媒介,这种媒介能够与学生的意愿及造型成果相一致,这些成果无论是特意的还是偶然为之的,随着研究的逐渐深入都会出现。每个学生,在交作业的时候,都要求递交一篇关于方法论的文章,非常简单但与每个人的研究内容相关。

图15 鲁昂博物馆

1.4 Particular experiences in two French schools

1.4.1 ATR teaching example established in 2008 at the ENSA in Normandy

A Point in Space

Goals

This course was designed for the 1st year students, which is crucial since it lays a foundation on issues related to representation at the ENSA in Normandy. It begins with a lecture, which raises some essentials stakes of the Western space. In fact, this course has been the introduction for the 2010 workshop at the Southeast University.

The goal is to provide a theoretical and practical basis on different types of representation, including drawing, by promoting:

·A confrontation between the scale of reality and scale of representation.
·An understanding of the passage between three and two-dimensional space and vice versa.
·Acquisition of historical and theoretical knowledge on the art and techniques of representation.
·An accuracy of observation and restitution in line with measures implemented.
·A freedom of imagination based on tools that favor personal expression.
·An exploration of new disciplinary, artistic or technical fields, focusing increasingly on the human body itself as a vehicle for the investigation of new spaces.

Teaching method

As part of the examples which are discussed, the lecture is followed by a visit to the Museum of Rouen(Figure 15), and students are invited to make some drawings based on a freely chosen painting, and then a first large model for the next course. The exercise typically lasts 5-6 weeks and alternates between two-dimensional and three-dimensional depiction. As they make headway in their work, the corrections are first collective and then highly individualized. Depending on their approach, each student is encouraged to gradually find a medium that is in line with his/her intentions and the plastic discoveries, whether intentional or accidental, which appear during his/her work progress. A methodological and simple text which accompanies the project is also required on the day of handing in.

教学案例

选取的教学案例有三种不同的代表。第一个案例,这个学生拥有非常好的技巧,凭借这些技巧,他能够以颠覆性的方式进行概念设计,并能以极其细腻、极其精巧的方式加以实现。第二个,与之相反,这个学生并没有接受过任何特殊教育,其绘画水平非常差,但他却有着很强的直觉,在色彩和材质上表现得很强,并能够以非常感性的方式将接受到的参照融入到研究中去。最后,第三个案例是前面两者,也就是前面两个学生的情况兼而有之,技术水平中等,但他在初始想法方面却能够超越前面两人,并敢于尝试一种与其自身肢体表演相联系的表现尺度,作为其表现工作。

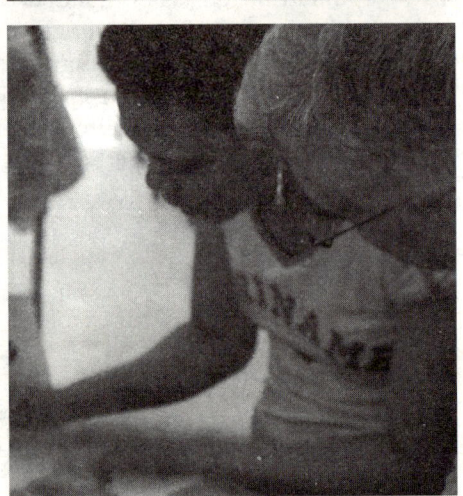

Examples

The examples chosen are three different cases: the first one is that of a student who has a very good technique on which he can rely in order to conceptualize drastically, while doing so in a precise and subtle manner. The second one, on the other hand, is that of a student who hasn't received any special training. Therefore, the quality of the design is poor, but the student made up for it with an intuitive approach, playing with the color and texture, and sensitively integrating the assigned references. The third example is that of a pair of students, who have a medium technical level but who have transcended their initial reflection and dared to work with a scale of representation related to the performance of their own body as a tool of representation.

案例 1

《倾倒的水壶静物》
（威廉·卡尔夫，1622—1693）

法国诺曼底国立高等建筑学院，2010
安东尼·弗雷

Still Life with Reversed Gourd
Willem Kalf, 1622—1693

ENSA of Normandy, 2010
Anthony Ferré

Nature Morte à la Gourde Renversée
Willen Kalf, 1622—1693

ENSA de Normandie, 2010
Anthony Ferré

威廉·卡尔夫（Willen Kalf）是佛拉芒地区最伟大的画家之一，他以其静物自然画闻名于世。该生在分析了这位画家的作品后，让我们参与到其观察过程中来："这些信息意味着不稳定，意味着静止，餐盘只是被水壶固定住，水果被吃掉了一半，而水壶也被打翻在桌上。 由此，我产生了从物理上塑造这种不稳定性的想法，为这一静物还原一个物理的自然现实。"通过制作一个盒子来表现其三维性，在盒子的内部放着一些物品，随后将这些物品染色。学生以同样的物品属性进行创作，但这样的属性"完全被混淆了"。"这是映像和真实事物之间本身属性的消失，是对属性本身所作的一场游戏。"

After having carefully deciphered the work of one of the greatest Flemish painters, famous for his paintings of still life, the student shares with us his observations: "These messages symbolize frailty and death - the plate is only held by the gourd, the fruits are half eaten and the bottle is turned upside down. Hence my idea of creating this physical frailty, a physical nature for this still life. " By restoring a three-dimensionality by creating a box within which objects are installed, and then painted, the student works on the nature of these objects which are "in complete disarray". "It's a loss of nature, a game of nature, between image and reality."

Après avoir décrypté attentivement l'œuvre d'un des plus grands peintres flamands, très connu pour ses tableaux de nature morte, l'étudiant nous fait part de ses observations : "Ces messages symbolisent la fragilité, la mort, l'assiette n'est retenue que par la gourde, les fruits sont consommés à moitié et la gourde est renversée. De là mon idée de créer physiquement cette fragilité, créer une nature physique à cette nature morte." En restituant une tridimensionnalité par la réalisation d'une boîte à l'intérieur de laquelle sont installés les objets, puis en les peignant, L'étudiant travaille sur la nature même de ces objets qui est alors "totalement chamboulée". "C'est une perte de nature propre, un jeu de nature, entre l'image et le réel."

《倾倒的水壶静物》
威廉·卡尔夫

安东尼·弗雷

案例 2

《在梅雷维尔公园乡村的桥上看尊老庙》
（于贝尔·罗贝尔，1733—1808）

法国诺曼底国立高等建筑学院，2010
塞巴斯蒂安·德内肖

学生首先制作了第一个模型，微型的，甚至有些笨拙，随后他用揉皱的纸制作了第二个模型。对于于贝尔·罗贝尔（Hubert Robert）的作品，他这样说道："这个作品让我想到了风一般的轻盈……我从中加入了自己对材质以及色彩的理解，让所有这一切变得更加抽象的同时最大限度地体现我的感觉……在渐变的色彩中，白色代表着能够确认身份的个人。彩纸塑造了一个世界，塑造了一个空间，能够让我们进入浅浮雕的内部。"这位学生天马行空地从西蒙·汉泰（Simon Hantaï）和罗伯尔·李曼（Robert Ryman）的作品中汲取灵感，他最后递交的作业体现了整个作业的连贯性，尤其是成果表现了演绎某种轻盈并"以其材质构建深度和空间"。

《在梅雷维尔公园乡村的桥上看尊老庙》
于贝尔·罗贝尔

塞巴斯蒂安·德内肖作品

The Temple of Filial Piety Seen from a Rustic Bridge in the Park Méréville
Hubert Robert, 1733—1808

ENSA of Normandy, 2010
Sébastien Denéchaud

Le Temple de la Piété Filiale vu d'un Pont Rustique, dans le Parc de Méréville
Hubert Robert, 1733—1808

ENSA de Normandie, 2010
Sébastien Denéchaud

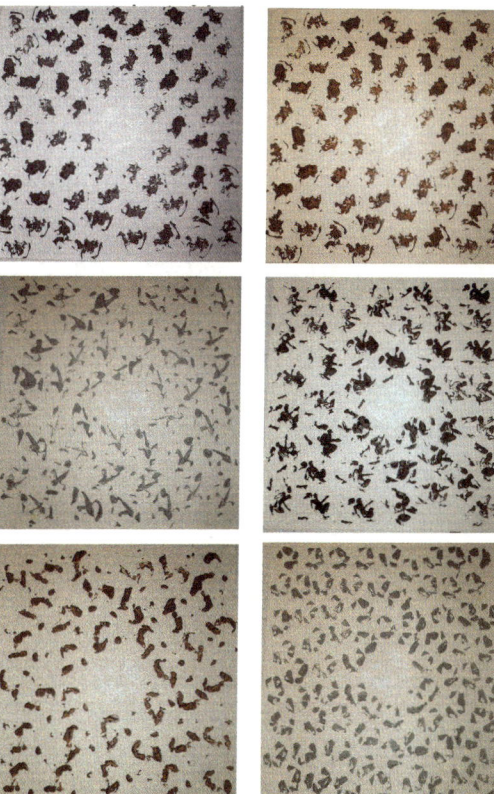

The student has built a first compact and rather clumsy model, then in a second step, a second model with crumpled paper. On Hubert Robert's painting, he said: "This work inspires me with lightness like a gentle breeze ... I inserted my own reading of materials and colors to transcribe most of my impressions, and make the work more abstract... In the midst of a gradient of color, white smears represent individuals whom we can identify. The colored papers create a world, a space that invites us inside this bas-relief." While being deliberately inspired by the works of Simon Hantai and Robert Ryman, his final report is a continuation of his work, particularly with works that have a certain lightness and "which by virtue of the material create a depth and space."

L'étudiant a construit une première maquette, plutôt compacte et malhabile, puis dans un deuxième temps, une seconde maquette avec des papiers froissés. Du tableau d'Hubert Robert, il dit : "Cette œuvre m'inspire de la légèreté comme un courant d'air... J'ai inséré ma propre lecture des matières et des couleurs afin de retranscrire au maximum mes impressions, et en rendant le tout plus abstrait… Au milieu d'un dégradé de couleur, des touches de blanc représentent les individus qui permettent de s'identifier. Les papiers de couleur créent un monde, un espace qui nous invite à l'intérieur de ce bas relief." Tout en s'inspirant délibérément des œuvres de Simon Hantaï et de Robert Ryman, son rendu final s'inscrit dans la continuité de son travail, notamment avec des réalisations qui retrouvent une certaine légèreté et "qui par l'effet de matière créent une profondeur et de l'espace."

案例 3

《屠杀无辜者》
（雅克·斯特拉，1596—1657）

法国诺曼底国立高等建筑学院，2010
蒂博·高第埃和格兰·德祖利

这两位学生从同一幅油画出发，先是各自研究，随后是共同研究。通过对这幅油画的分析，两位学生提出了两种情况下完全截然不同的富有戏剧性的设计模型。其中的一个模型，在现代的环境下构建了一幅景象，宫殿和宗教建筑已被一栋住宅和一座水塔所取代。他这样写道："为此，我做了第一个模型，它表现了我所想要重现的想法，象征所有参与者（演员）……因为在我看来，这是最主要的主题，而且这比场景更为重要，因此它能够在任何地方展开。在经过一番思索之后，我又仔细斟酌布景，斟酌我所想要传递的信息并为我的想法赋予真实的一面：让观众成为演员……"

这两位学生因此也成为"演员"，并进行行为绘画，"过程是受到了油画主题的启发，一方面保留了油画中的灰度，同时表现了场景的暴力面，保留了其精髓。"

《屠杀无辜者》
雅克·斯特拉

蒂博·高第埃
格兰德祖利

Massacre of the Innocent
Jacques Stella, 1596—1657

ENSA of Normandy, 2010
Thibauld Cordier and Glenn Desury

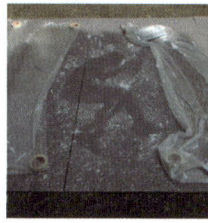

The two students worked from the same painting, at first separately, then together. The analysis of the painting has led the students to offer two distinct and highly theatrical models in both cases. One of them set the action on stage in a modern environment and the palace and the religious building were replaced with a building and a water tower. Thus he wrote: "For that, I made the first model, which expresses the idea that I wanted to transcribe and symbolizes all the actors ... because in my opinion, it is the main subject and it is more important than the place itself ... It can take place anywhere. After careful consideration, I thought of the sets, I wanted to transfer the message and give a real feel to my idea by considering the viewer as an actor ... "

Both students have thus become "actors" and have taken part in the painting action "with a method inspired by the theme, not only keeping the gray of the painting, but also expressing the violence of the scene. Keeping the essential."

1.4.2 巴黎玛拉盖国立高等建筑学院 于2002年实施的表现艺术和方法教学案例

《从历史的深度到当代的平面性》

教学目标

巴黎玛拉盖国立高等建筑学院的这种教学方法针对的是2002—2006年在校期间的三年级学生。这种教学法也曾于2008年在法国诺曼底国立高等建筑学院及2010年在中国南京东南大学的教学活动中施行过。

其目的在于，当学生研究现代艺术，尤其是当代艺术的时候，让他们不要停滞于当今的艺术现状，从而能够对艺术的本源提出批评性的提问。对历史编年的好奇，能对于未来有着持久的展望的态度。

这种教学法基于对寓意画像艺术以及编年史的一些知识和标志，同时在介绍过程中采用闪回（倒叙）的方式，通过广泛调查的方式将当代艺术置于历史的长河中，从而让学生对当代艺术进行解读或重新解读。在这六门课中，每一门课都涉及一个确定的主题。这些主题是我们逐渐看完整个系列作品后定义的。这些主题可以是一个时代、一场运动或是我们至今仍关注的某些艺术家。

教学方法

这些理论课程一开始就通过建议一个练习主题来展开，学生根据这个主题（在8天内）做出一个具体的作品并撰写一篇简短的文章（1~2页），既作为对课程的回应，也作为对以后研究步骤的解释说明。主题都与当天的主题以及之后的问题和讨论相关。它们的目的在于，让学生形成批判性眼光，当他们在二维或三维的空间内，或者处理这些空间时，让他们更深刻地认识到其重要性，从而使得他们能更加自如地在随后的工作室的学习中实现他们的作品。

对实践的集体讲评，对大多数学生来说，将是非常具有启发意义的，将会对先前的课程形成更广泛的诠释和更好的理解。

教学案例

所有的教学案例都是按照其质量来选择的，从精心制作的各个步骤，到对某个想法的执着追求，而且与其恰当的实现方式是一致的。它们对应着各自不同的练习，我们可以在和每份作业一起交上来的简短文章开头找到这些练习。在最初的几个案例中，4个案例的名称稍微有所不同，对应的是各个不同的学年。在最后几个案例中，学生们依据同一个题目，做出各自的作品。每一次我们都能发现交上来的作品各不相同，这种多样性也体现在南京的学生交上来的作业中。

1.4.2 ATR teaching example put up in 2002 at the ENSA Paris-Malaquais

From the Historical Depth to the Contemporary Plan

Goals

At ENSA Paris-Malaquais, this course was generally designed for 3rd year students from 2002 to 2006. It was resumed in 2008 at the ENSA in Normandy and recently adapted for the workshop at the Southeast University of Nanjing in 2010.

The aim is to teach students who look at modern art, and especially contemporary art, not to stop at the present day layer, but rather to develop a critical interrogation of the origin. Being curious about chronology supports a sustainable forward-looking attitude towards the future.

Based on some knowledge and iconological and chronological benchmarks, this course, through the use of flashbacks, also offers students a reading or rereading of contemporary art by placing it in a widen investigation of the historical field. Each of the six courses addresses a specific theme that is gradually defined in relation to a group of works, whether from a period, a movement or a few artists on which we focus our attention.

Teaching method

These courses are immediately followed by a proposed exercise topic from which the student produces a concrete implementation (in 8 days) as well as a short paper (1-2 pages) which is related to the course and explains the approach for the choice of answer. The subjects are directly related to the theme of the day and to questions or discussions that emerged from it. They are intended to stimulate the students' critical perception and lead them to be more aware of issues when working in or with a two or three dimensional space. This will help them tackle with greater ease the manipulations they carry out in their project studio.

The collective feedback on the exercise is a moment which reveals the multiplicity of responses and provides a broader interpretation and understanding of the previous class.

Examples

All examples have been chosen for the quality of the approach to the relevance of an idea, and its consistency with its materialization. They respond to different exercises whose titles are to be found at the heading of the texts accompanying each record. In the first case, four examples have slightly different titles and correspond to different academic years. In the latter cases, the works were made after a single title. Every time, there is a variety of answers, this diversity is also present in the work of students in Nanjing.

案例 1

白上之白,线为之线

法国诺曼底国立高等建筑学院,2010
纪尧姆·雅盖(硕士一年级)

这一作品来自于研究建筑的想法,以及在阅读主题时所强烈感受到的直觉。当我看到"建筑"、"白色"和"线条"这些词的时候,宏伟建筑物的形象就扑面而至,我知道我可以做这个主题。因此,凭借着直觉,在找到了韩国首尔宏伟建筑物的强烈形象之后,我开始拼装这些白色的空间,并留出交通空间的脉络,从而突出构建这一形象的主要线条。

我的目的在于,制作一个物体且有意识地提出本质问题。事实上,我想要做的是这样一个吸引人的物体——能够代表我对此类建筑批判性思考的物体,能够突出此类正交轮廓线非人性的一面。

White on White, Line for Line

ENSA of Normandy, 2010
Guillaume Jacquet

Blanc sur Blanc, Trait pour Trait

ENSA de Normandie, 2010
Guillaume Jacquet

This work is the result of a willingness to reflect upon architecture and a strong intuition felt on the reading of the topic. When I came across the words "architecture", "white" and "line", the image of large sets came to me — I knew I could work on this subject. Thus, intuitively, having found a very strong image of large sets in Seoul, I began to mount white volumes and dig circulating veins to bring out the main lines that structure the image.

My goal was to create an object and put forward intentionally subjective questions. In fact, I wanted to make something appealing — it is deliberately the image of my critical reflection upon this type of architecture, which emphasizes the non-human aspect of this orthogonal layout.

Ce travail est l'issue d'une volonté de traiter l'architecture et d'une intuition forte ressentie à la lecture du sujet. Lorsque j'ai croisé les mots "architecture", "blanc" et "trait", l'image des grands ensembles m'est venue, je savais que je pouvais travailler sur ce sujet. Ainsi, de manière intuitive, après avoir trouvé une image très forte de grands ensembles à Séoul, j'ai commencé à monter des volumes blancs et creuser les veines de circulation principales pour faire ressortir les lignes qui structurent l'image.

Mon objectif était de réaliser un objet qui questionne et volontairement subjectif. En effet, je voulais réaliser un objet qui interpelle, qui se veut volontairement l'image de ma réflexion critique sur ce type d'architecture, qui souligne l'aspect non-humain de ce tracé orthogonal.

案例 2

白上之白，形象之形

法国巴黎玛拉盖国立高等建筑学院，2002
皮埃尔·伯纳姆尔

画家用的网格叠放在一面镜子上，当中留下大约1厘米的间隔。网格和镜子通过框架支持并固定住。这个框架是个旧的镀金框架，因年代久远而产生了旧的光泽。

金属网格表面的圆圈，能够让光线穿过表面，从而在镜子中重新形成网格映像的"形象之形象"。

网格对称的几何构造，被完全浸入了画面中，从而在镜子中再次映射出能够看到的一面，从而达到"白上之白，形象之形"。

框架成了作品不可分离的一部分，如同其与墙壁和颜色一般不可分离，它的角色从容器转向了容纳物本身。事实上如果这个作品挂在墙上，但采用的是其他的颜色，而不是网格，那该作品的框架与作品不会如此不可分离：它更表现得像是墙壁上网格画面的支撑物。

在这个练习中，我希望展示的是，对构成的感知，既与其本身的构成部分密切相关，也与周围的环境因素密不可分。

White on White, Line for Line

ENSA Paris-Malaquais, 2002
Pierre Bonnamour

A painted grid is superimposed onto a mirror. The space between them is about 1cm. The grid and the mirror are kept and attached via a frame. This frame is ancient, gilded and patinated by the time.
The circles on the surface of the metal grid, let the light go through the surface and creat a reproduction "feature for feature" of the image of this grid in the mirror.
The symmetrical construction of the geometry of the grid was completely soaked in paint. It provides, on the mirror, the reproduction of the face and then overlaid "White on white, line for line".
The framework became inseparable from the object, as it was inseparable from the wall and its hue. It changed its status from a container to contents. In fact, if the same object was hung onto a wall with a different color from the grid, its framework would not come off the same way: instead of being itself as part of a composition, it would appear as a support for a painting grid on a wall.
 In this exercise, I wanted to demonstrate that the perception of a composition is related to both its components and its environment.

Blanc sur Blanc, Traits pour Traits

ENSA Paris-Malaquais, 2002
Pierre Bonnamour

Une grille de peintre est superposée avec un espacement d'environ 1cm à un miroir. La grille et le miroir sont tenus et fixés par l'intermédiaire d'un cadre. Ce cadre est un cadre ancien doré, patiné par le temps.
Les cercles, à la surface de la grille métallique, laissent la lumière traverser la surface permettant une reproduction "traits pour traits" de l'image de cette grille sur le miroir.
La construction symétrique de la géométrie de la grille associée au fait qu'elle ait été trempée entièrement dans la peinture permet d'obtenir sur le miroir une reproduction de la face visible et donc de superposer "Blanc sur blanc, traits pour traits".
Le cadre devient indissociable de l'objet qu'il contient en même temps qu'il est indissociable du mur et de sa teinte, Il passe du statut de contenant à celui de contenu, en effet, si ce même objet est apposé à un mur d'une autre couleur que la grille, son cadre ne se détacherait pas de la même manière : au lieu de s'affirmer en tant qu'élément d'une composition, il apparaîtrait comme support d'une grille de peinture sur un mur.
Dans cet exercice, j'ai souhaité démontrer que la perception d'une composition est liée tant à ses éléments constitutifs qu'aux éléments environnementaux.

案例 3

平面之白，线条之线

法国巴黎玛拉盖国立高等建筑学院，2003
雅德·于斯特

"平面之白，线条之线"，这些限定了主题的表述方法，都非常直观：我们在此谈论的是颜色、形状和表面；然而，正是这种感性对我们提出了回应这一既定主题的要求。因此，正是拥有其内在的属性，使我们在思考之前就将之看作一个物体而不是一种观点。

正是基于这种考虑，我首先试图寻求的是一种尽可能最"客观"、最"真实"的对物体的表现（逐字诠释）。我想说的"真实"，是通过物体以其独特的外形来回归这个主题，而不需要作任何解释。

我的第二个想法仅仅是为了引入艺术参照物。事实上，这个主题让我想到了让-皮埃尔·雷诺的作品。这些作品围绕着用黑色线条分隔出来的白色彩陶方块（这两种表述方法意思相同，只是"线条"这种说法比"线"更为考究，更加细腻。）。

我最终做出的作业只是融合了这两种表现方式的要求。

我所追求的极端真实，可以通过顷倒在平坦表面（平面）上的白色油漆加以体现。然而，艺术参照物是通过选择一个现代城市的平面图来象征化的，如同国际象棋的棋盘，形成了如同雷诺作品一样的格子。这是比东京的地图简单的格状图。这个城市里街道没有名称。这些无名的街道在图上成了线条。

油漆倒在由线构成的平面上，于是，白色的油漆沿着平面将整个平面占据，并抹去了上面重要的线。

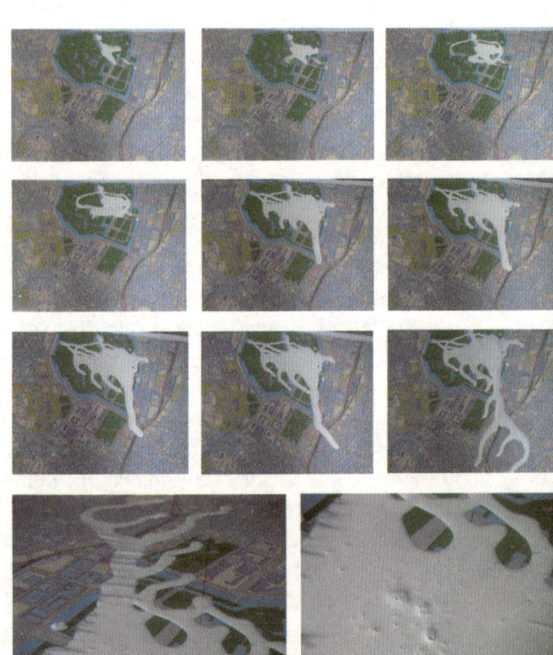

White on Map, Line Drawing

ENSA Paris-Malaquais, 2003
Jade Juste

"White on map, line drawing" : the terms that define the subject are exclusively visual: we're talking about color, shape and surface. Yet it is through the sensory paper that we are asked to respond to this topic. Thus, it has inherent qualities that lead us, before even thinking, to perceive it as something more than an idea.

From this consideration, I first attempted to seek a representation as "realistic", as "real" as possible of the subject (a literal interpretation). By "real", I mean in the sense that the object will return to the topic through its mere form, without any need for explanation.

My second idea is just an introduction of artistic references. Indeed, the topic reminds me of the work of Jean-Pierre Raynaud around the white square tiles separated by black lines (the two terms mean the same thing, except that "line" is more formal and finer than "drawing").

My end result is simply the union of these two constraints of representation.

The real exaggeration that I'm seeking has been accomplished with the white paint poured onto a flat surface (a map), while the referent art is symbolized by the choice of a modern city plan, in the shape of a checkerboard, forming a grid in the manner of Raynaud tiles. This map is a simpler grid than that of Tokyo, a city where the streets have no name, where the anonymous streets on the map become just lines.

The painting will be placed on a related compound, so that the white runs down the plan, covers it and erases the outline.

Blanc au Plan, Trait à la Ligne

ENSA Paris-Malaquais, 2003
Jade Juste

"Blanc au plan, trait à la ligne", les termes qui délimitent le sujet sont exclusivement visuels : on parle ici de couleur, de forme, de surface ; or c'est par le sensible qu'il nous est demandé de répondre à cet énoncé. Ainsi, ce dernier possède des qualités intrinsèques qui nous mènent, avant réflexion, à le penser comme une chose plus que comme une idée.

A partir de cette considération, j'ai tout d'abord pris le parti de rechercher une représentation la plus "réelle", la plus "vraie" possible du sujet (une interprétation mot à mot) ; j'entends "vraie" dans le sens où l'objet devra ramener à l'énoncé par sa seule forme, sans qu'il y ait besoin d'explication.

Mon second parti-pris n'est autre que l'introduction de références artistiques. En effet, l'énoncé évoque pour moi l'œuvre de Jean-Pierre Raynaud autour du carré de faïence blanc séparé par des lignes (les deux termes ont le même sens, à cela près que "ligne" est plus soutenue, plus fine, que "trait") noires.

Ma réalisation finale n'est autre que la réunion de ces deux contraintes de représentation.

Le réel outrancier recherché s'est vu concrétisé par la peinture blanche versée sur une surface plane (un plan), tandis que le référent artistique est symbolisé par le choix d'un plan de ville moderne, en damier, formant un quadrillage à la manière des carreaux de Raynaud. Ce plan tient d'autant plus du simple quadrillage qu'il s'agit de Tokyo, ville dans laquelle les rues n'ont pas de nom, ces rues anonymes du plan deviennent simplement des lignes.

La peinture sera versée sur un plan composé de traits, ainsi, le blanc coule le long du plan, le recouvre, en efface les grandes lignes.

案例 4

白（空白）上之白

法国巴黎玛拉盖国立高等建筑学院，2006
马利翁·拉卡

存在着各种各样的白（空白），包罗万象的"白（空白）"，等待成形的白色，洁白无暇的白色。

同样，也存在着能够隐蔽的白色，能够乔装打扮的白色，裹尸布般的白色。

但有没有"确定"的这如同幽灵般的界定，融合并区分了这"白（空白）上之白"？

缺少上部覆盖物的"空白"；白布在等待着填补缺少下部物体的"空白"；床单在等待着身体。

这就是敞开和遮蔽之间面对面、背靠背的痕迹。

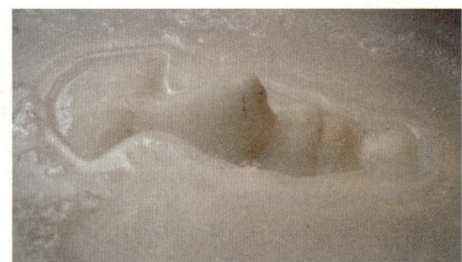
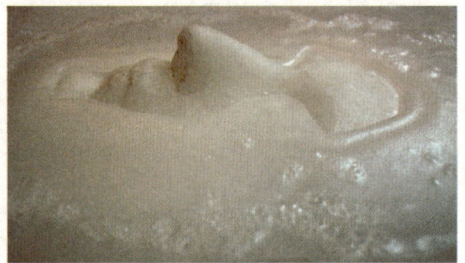

White on White

ENSA Paris-Malaquais, 2006
Marion Lacas

Blanc sur Blanc

ENSA Paris-Malaquais, 2006
Marion Lacas

There is the "white" that remained white while waiting for the form, the virginal white.
And then there's the "white" that covers the white mask, white shroud.
But what about the "on", the limit that ghostly unites and distinguishes these two absences, "white on white"?
"BLANK" in the absence of an upper; canvas waiting to appear on "WHITE" lack of a below; cloth awaiting body.
Here is the trace of this face to face with the back to back between the open and covered.

Il y a le "blanc" qui reçoit, le blanc en attente de forme, le blanc virginal.
Et il y a le "blanc" qui couvre, le blanc qui masque, le blanc du linceul.
Mais qu'en est-il du "sur", de cette limite fantomatique qui unit et distingue ces deux absences : "blanc sur blanc"?
"BLANC" de l'absence d'un dessus; toile en attente de figure sur "BLANC" de l'absence d'un dessous; drap en attente de corps.
Voilà la trace de ce face à face, de ce dos à dos entre l'ouvert et le couvert.

案例 5

形象之轮廓/轮廓之形象

法国巴黎玛拉盖国立高等建筑学院，2006
马利翁·拉卡

    相似应如同儿子和父亲的相似，这种相似性往往调和了巨大的物理差异，但并不取决于任何东西，而取决于神态，正如同当今的画家们所说的那样：一旦看到儿子就能想到父亲，但比较这两者，又发现其实完全不同，然而，不知出于何种原因，有种神秘的力量维持了两者之间的紧密联系。
                  ——彼特拉克《与薄伽丘书信》

    形象之轮廓在此真正将我们区分了出来，然而，通过对我们镜像的对比，我们不再看到自己的形象，而是一个形象投射到另一个形象上的形象：轮廓之形象，一个有着永存的紧密联系的画面。

Silent Face / Face Drawing

ENSA Paris-Malaquais, 2006
Marion Lacas

The likeness must be similar to that of a son to his father, which often accepts a large physical difference and which is worthless, one tune, as the painters of today say: once that we see the son, the father comes to mind, the comparison between the two shows they are very different, and a mysterious thing that I do not understand yet maintains the approximation.
—Letter to Boccaccio, Petrarch

The features appearing here are what distinguish us from one another, but by comparing our reflections we no longer see our own figures, but only that through which we are related to one another: the drawing of a face, the image of a unfailing link.

Tait de Figure/ Figure de Trait

ENSA Paris-Malaquais, 2006
Marion Lacas

La ressemblance doit être analogue à celle d'un fils à son père, qui s'accommode souvent d'une grande différence physique et qui tient à rien, à un air, comme disent les peintres d'aujourd'hui : aussitôt qu'on voit le fils le père revient en mémoire, la comparaison entre les deux les montre tout différents, et pourtant un mystérieux je ne sais quoi maintient le rapprochement.
—Epître à Boccace, Pétrarque

Le trait de figure est ce qui ici nous distingue, mais par la confrontation de nos reflets nous ne voyons plus nos propres figures mais ce qui dans l'une nous renvoie à l'autre: une figure de trait, l'image d'un indéfectible lien.

案例 6

形象之轮廓线/轮廓线之形象

法国巴黎玛拉盖国立高等建筑学院，2003
桑德丽娜·勒米尔

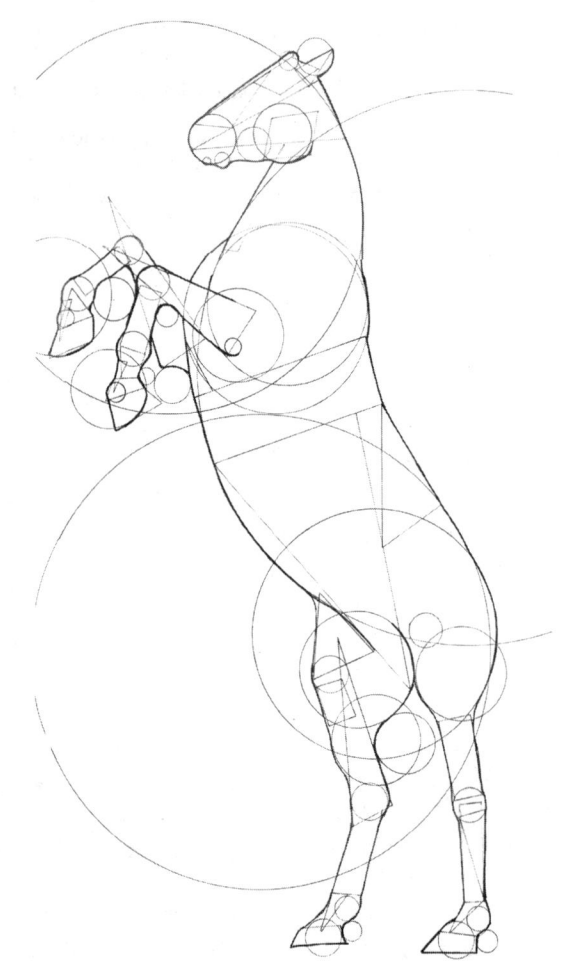

　　针对这一主题，我从多个几何形象出发，也就是所谓的构建图形，以获得一根轮廓线，然后是另一根轮廓线，之后是其他的轮廓线。其结果就是索米尔黑骑士的坐骑形象：腾跃的马。"形象"一词在拉丁文化中意味着外形，这里就是这种情况：它们形成并决定了轮廓线，并通过矢量工具赋予其生命和动感，构成一个形象。画出的马的图形，结构良好，将其与未经处理的背景区分开来。然而，在马的外围，在突出平面几何图形的同时，我让这个形象与背景融合。这样似乎显得更加稳定，背部弯曲，四肢划出了先前的动作。

Face Stroke and Face Drawing

ENSA Paris-Malaquais, 2003
Sandrine Lemire

Figure de Trait et Traits de Figure

ENSA Paris-Malaquais, 2003
Sandrine Lemire

For this work, I begin with several geometric figures, saying construction for a line, then another and yet another. The result is an equestrian with the Black Frame of Saumur: by Courbette. The original Latin "form" signifies figures, and this is what we have here: they form and determine the line and then give life and movement to another figure with a "tool" that serves as a vector. The equestrian is shown fully structured, and detached from the untreated bottom. However, by exacerbating the geometric figures on the surface, outside of the contour of the horse, I put the figure into the background. Then it seems to acquire more balance with its back bending back and its limbing describe the traces of a previous movement.

Pour ce sujet, je pars de plusieurs figures géométriques, dites de construction, pour obtenir un trait, puis un suivant et encore un autre. Le résultat est une figure équestre du Cadre Noir de Saumur: la Courbette. L'origine latine de figure signifie forme, et ici c'est le cas : elles forment, déterminent le trait et ainsi donnent vie et mouvement à une autre figure grâce à un "outil" qui sert de vecteur, le trait. La figure équestre représentée est entièrement structurée, ce qui la détache du fond non traité. Cependant, en exacerbant les figures géométriques dans le plan, bien en dehors du contour du cheval, je l'inscris dans le fond. Il semble alors prendre plus d'aplomb, sons dos se courbe, ses membres décrivent les traces d'un mouvement précédent.

2 东南大学建筑学院教学活动

2.1 教学活动的准备

当然,正是基于上述的观察和经历,我们才提出了教学的建议,但需要记住的是,我们面对的是中国学生。毋庸置疑,王盈先生起到了举足轻重的桥梁作用,因为他早先在北京接受了建筑学的教育,然后又再次在巴黎学习。除了他的翻译能力外,他曾经在北京和巴黎学习的经历,敏锐地理解对话双方的所有需求。

首先,我们尽力理解东南大学建筑学院的期望所在。我们在前期通过与中国同行进行长期的沟通作了大量的准备工作,为的是更好地明确我们中方合作伙伴的期望,并使之与我们的能力和手头可用的各种素材结合起来,而且我们知道我们要和一个大的团队在有限的时间内一起合作。

2.2 日程安排及进展

在教学活动的预定日程和实际安排之间,我们实际上有一定的差别,尤其是将理论课程分割开来,以便学生能更好地理解领会。简单的描述每日的教学活动,似乎是让人了解我们教育理念的最好的方法。东南大学随后的期望是,在一段时间之后,能够逐渐的总结出主要的教学方法。

第零天 /8月23日/

抵达东南大学。教学活动简介,明确校方的期待以及我们的建议和最后调整。并翻阅了几本珍贵的书,尤其是杨廷宝先生从法国带回的那些书。

第一次的会议表现出我们大家都已经作好了一起工作的准备。主要的议题是:我是选择对三十名学生的一个班进行教学指导还是对整个年级的所有学生进行指导。最后我们选择通过一个"试点"班的模式对整体学生进行教学指导。这是一个雄心勃勃的项目需要良好的沟通,但这一切看起来似乎我们都已经做好了准备。

第一天 /8月24日/

上午,第一个讲座(两个半小时左右),和原先的安排一致。该讲座给所有学生带来初步的基本信息和知识,或是作为新学到的或是回顾已学到的。

为了使教学活动顺利进行,需要奠定初步的理论基础,而且这些基础同样也能够在欧洲文化和中国文化之间搭建一些沟通的桥梁。这个讲座以布置第一个练习为结尾,这个练习以法国作家、诗人、哲学家和认识论专家保罗·瓦莱利(Paul Valéry)*的一句话出发:"对于人来说,最具深度的是其皮肤。"

* Paul Valéry (1871-1945), *L'Idée fixe ou Deux Hommes à la mer*, 1932.

2 Workshop at the School of Architecture of SEU

2.1 Preparation of the Nanjing workshop

Of course we established propositions according to the observations and experiences previously mentioned, and kept in mind the fact that we are talking to Chinese students. Wang Ying of course acted as a key intermediary, as he received architect training, first in Beijing, and then in Paris. In addition to his competences as a translator, his experience as a former student in Beijing and in Paris, gives him the ability and subtlety to understand both the explicit and implicit demands of our interlocutors.
Firstly, we tried to understand what were the requirements of the School of Architecture and the Southeast University. We deliberately did an important work of preparation, based on a permanent contact with our Chinese partners in order to define accurately their expectations, inform them about our skills and the material capacity that we could put at their disposal, knowing that we would have to work with large staff in a limited amount of time.

2.2 Day by day diary of events

Between the initial forecasted and the actual workshop reality was, many changes have been made. For instance, some theoretical classes have been fractioned in order to improve the students' understanding. Describing the day by day sequence of events in this first publication seems to be the best way to present our pedagogical philosophy. We are aware of the fact that the main issue of Southeast University is to be able to extract from it methodological aspects for the future.

D0/ 23rd of August/

Arrival at Southeast University. Presentation of the workshop, last adjustments of the expectations and of our proposals. Visit of the buildings, including the library and the discovery of some valuable books brought from France by Yang Tingbao.
The first meeting showed that we were ready to work together, and the main issue was whether to choose to work with one single group of thirty students, or to work with the whole of the year. By running tests with a pilot group of students we finally decided to go for the most collective option. It was an ambitious project, which would require good intermediaries, but we strongly felt that we were ready for this.

D1/ 24th of August/

In the morning, the first lecture (about 2h30) corresponded to what was expected, and was meant to give the students a first basis of information and knowledge, learnt for the first time of reviewed.
For a successful workshop, the theoretical basis was essential and we also established interesting links between the Chinese and European culture.The conference ended with the introduction of exercise 1, based on a quote by Paul Valery, a French writer, poet, philosopher, and epistemologist :"On a man, the skin is the deepest".
* Paul Valéry (1871-1945), *L'Idée fixe ou Deux Hommes à la mer*, 1932.
Then, we asked the students to come up with an appropriate proposal for the exercise: "Nothing is deeper than a surface".

然后我们让学生们以一句更合适的、接近的话来思考这个练习："没有什么比一个表面更具深度。"

这天下午被分为了两个部分：
· 和我们称为"试点"的小组开会。
· 作一次巡视，随后了解每个组的情况，与学生及其各自的指导教师进行沟通。这也使我们有机会参观他们的工作室。晚上和老师进行会谈，在交流结束的时候，我们决定在第二天进行一个简短的协调会（约40分钟）以补充和丰富前一天的内容。

第二天/8月25日/

上午，我们上了前一天决定的补充课程。实际上这节课非常短，持续大约40分钟。它继续了第一次讲座的内容，并为学生们打开更广的探索领域。

这次的授课，具有很强的针对性，并引用了前一天已经介绍过的几个艺术作品，但这次是在另外一种语境中，以一次展览为背景。可能正是这样具有决定意义的授课，使得学生们摆脱了束缚。

下午，我首先和试点小组在一起，随后，其他小组愿意参与的学生，也加入了我们。

晚上，我们和教师进行了第二次座谈。在座谈期间，我向他们展示了法国学生们的作品。这里要讨论的是第一次练习的评判标准。此外，我们一起明确了几个决定，并确认了几个教学选择：
· 每天都进行一次讲座，但更简短
· 要求学生就他们的主题写出三个

关键词，以帮助他们更好地勾勒他们自己的想法，并检验其是否坚持了这个想法。

· 系主任的介入，更好地解释了教学目的，并明确鼓励学生要有更多的自主性。曾琼先生的这次介入非常重要。

第三天/8月26日/

上午，第二个讲座（一个半小时左右）。这个讲座，和计划中预定的相比，更加简短，而且只进行原计划讲座的第一个部分。

下午，点评第一次练习。所有学生都参加。在很快地巡视一周之后，

The afternoon was divided into two:
· a meeting with the so-called "pilot" group. going through each group in order to establish contact between the students and their teachers. We also had the opportunity to visit the workshops. In the evening, a meeting was held with the teachers, and at the end we took the decision to offer a short intervention (40mins) for the following day, in order to complete and enrich the remarks of the previous day.

D2/ 25th of August/

In the morning We had the extra class which had been set on the previous day. It was indeed quite short and lasted for about 40 minutes. It mainly carried on with the preparation work started on the previous day, and opened a bit more the students' field of investigation.
This class, which was very precise and reviewed some artists' works seen on the previous day, but this time in a different context — an exhibition — allowed the students to express themselves more freely.
In the afternoon I first stayed with the pilot group, but then students from other groups joined us.
In the evening there was another teachers' meeting. I showed some works made by the French students. We had to discuss the grading criteria for the first exercise. In addition, we confirmed some decisions and specified some pedagogical choices:
· have regular meetings everyday, but make the meetings shorter.
· ask the students to write down three key words about their project, in order for them to have a better idea of what they wish to do, and to make sure they stick to it.
· The department director made a

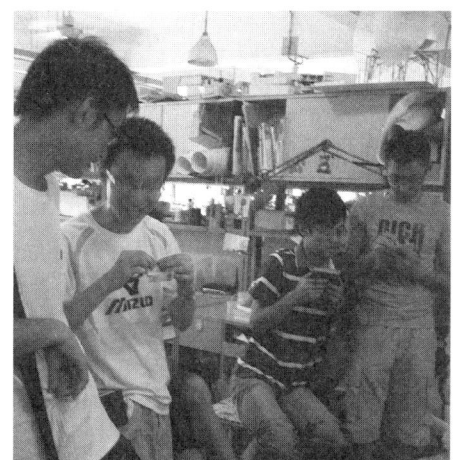

speech to explain the issues and strongly encourage the students to take some more liberty. Mr. Zeng Qiong's speech was very useful.

D3/ 26th of August/

Morning: The second conference (more or less 1h30). Shorter than what had been originally planned, only went through the first part.
Afternoon: Feedback on the first exercise. All the students were present. After a quick look, 30 works were picked out according to some predefined criteria (the concept's pertinence, the work's quality, coherence, poetic aspect...). We mainly picked the best projects; but not only. They were

我们根据预先设定的评判标准（设计理念恰当、作品的质量、前后一致性、表现的诗意……）选出三十多个作品。我们选出的是最好的案例，但也并非完全如此。随后，通过与相关学生或学生小组交流，集体点评这些作品。

在点评结束的时候，我们给出第二个练习的主题，该练习与上午的理论课相关，并要求学生后天交上此作业。这次引用的是巴黎玛拉盖国立高等建筑学院（ENSAPM）的建筑师兼教师夏维埃·法布尔*（Xavier Fabre）的一句话："所有建筑都是对运动的框景（取景）。"

*夏维埃·法布尔（Xavier Fabre），《Repères》杂志18期（瓦勒·德·马恩舞蹈发展中心及舞蹈双年会出版），2006.

第四天/8月 27 日/

上午继续进行第二个讲座的后半部分（约40分钟），相对最初的计划真正做出修改和简化。之后，介绍了之

前几天和学生交流过程中提及的一些艺术家。

下午，针对试点小组的学生以及希望展示自己练习的学生进行讲解、再次定位和个别辅导。

第五天/8月 28日/

上午，第三个讲座——绘画时光（约两个半小时），分为三个部分。这次介绍按照最初拟定的提纲进行。

下午，按照前次练习点评的方式进行第二次练习的点评，并介绍第三次练习，对于这次练习，不给予任何特殊指令，只要和每个人的研究选择一致即可，当然，这也需要与整个教学活动的大主题保持一定联系。学生很自由，如有必要，可以发展或延续之前的两次练习，将它们进行总结，或是与此相反，采用一个全新的想法。

第六天/ 8月29日/

星期日休息

then collectively discussed and feedback was given within an exchange with the concerned student or group of students.

Following our collective discussion, we moved on to exercise n.2 based on that morning's theoretical class with a quote by Xavier Fabre*, professor at the ENSAPM: "any architecture is a frame of movement".

*Xavier Fabre, *Repères*, n°18, published by the Center of the Choreographic Development and the Biennale of Dance in Val de Marne, 2006.

D4/ 27th of August/

Morning: Second part of the second lecture (more or less 40mins). Quite altered and simplified from the original plan. Then some artists mentioned on the previous days during the exchanges with the students were introduced.

Afternoon: explanation, reframing, and beginning of individual feedback for the students of the pilot group and other students who wanted to show their work.

D5/ 28th of August
Morning : The third lecture—*The time of painting* (2h30 more or less) with a threefold presentation. This presentation followed the original schedule.

Afternoon : The second exercise was discussed and reviewed following to the same criteria as for the first exercise. Presentation of exercise n°3, for which no special instructions were given, except for the fact that each student had to remain coherent with his/her initial investigation choice, and with the theme of this workshop. The student is free and can develop the work done on the two other exercises already done, do a synthesis, or end up with a new proposition.

D6/29th of August
Sunday rest

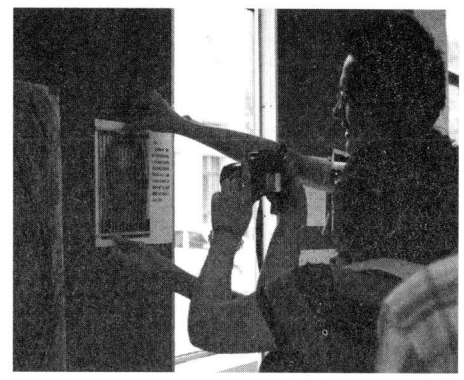

D7/ 30th of August

Morning: The last lecture introduced the artists mentioned during feedbacks or answering students' questions. Then had a presentation of some examples of European students' works.

Afternoon: Students worked to finish the third exercise. Then we did the last feedback with the pilot group and the students who wished to be involved. The next meeting is the presentation of the finished work, the following day at 2 P.M.

第七天/ 8月30日/

上午，最后一个讲座，介绍点评过程中提到的或是有些学生问到的艺术家作品，随后是精选的欧洲学生的几个作品。

下午，学生习作，准备第三次练习的最后交图，而我们则对试点小组或是希望参加的学生进行最后一次辅导。下一次碰面就是第二天14：00交最后作业的时间。

第八天/ 8月31日/

上午，所有教师开会，这次展现的首先是巴黎玛拉盖国立高等建筑学院以及诺曼底国立高等建筑学院所采用的方法论，其次是欧洲学生作品中非常具体的几个案例。

第九天/ 9月1日/

上午，我们和希望来的学生见了面，但主要谈论的是展示和表现的

D8/ 31st of August

Morning dedicated to a meeting with all the teachers, in order to to show other examples of European students' works, and some methodological aspects designed for ENSA Paris-Malaquais and ENSA of Normandy.

D9/ 1st of September

Morning: We met with any students who wished to, but mainly to answer some questions regarding their presentations. Afternoon: The final deadline was at 2P.M. The students were asked to make the presentation as clear cut as possible, emphasising their last work, in order for us to understand clearly their approach to their work(s). Some students presented all of their projects from the workshop.
For this last feedback, a bit more than forty works were selected. The selection was made by all the teachers, in order to select all the works which seemed to be the best and most pertinent.

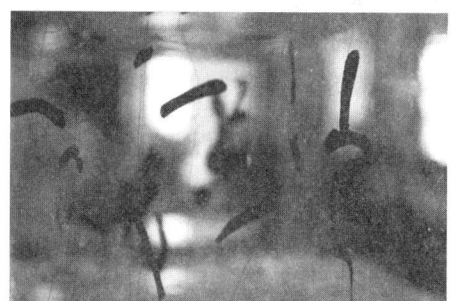

问题。

下午，最后交图时间为14:00。学生或是学生小组被要求进行尽可能严谨的布展，使其作品更具价值，以便我们就能很容易地理解他们的方法及作品之间的关系。有些学生展示了在教学活动的过程中所有的作品。

对于最后一次点评，我们选择的稍多一点的作品，有四十多个，而且这一次，真正是所有的教师都参与到评选中来，以选择我们认为真正最好最恰当的作品。

2.3 定制课程的内容

在这一章，我们将介绍这些讲座的原先预定的内容，每个讲座被预先分为两个部分。随后，遵循此前日程进展的逻辑，我们会指明此次在东南大学教学活动中的调整内容。

2.3.1 课程一：空间中的一点

第一课至关重要。它能够启发灵感，并通过分析西方的二维表现方法，奠定整个教学活动关键的理论基础。当然，优先要分析的是意大利文艺复兴时期创造的"中央透视法"，同时也让学生做好准备，更好地理解某些演变，并且用更警觉更谨慎的目光，去审视当今的各种表现方式。

讲座提纲

第一部分

（1）重要里程碑
（2）从乔托（Giotto）到马蒂斯（Matisse）
（3）第一次文艺复兴
（4）第二次文艺复兴

第二部分

（5）作品中的绘画 *(2001年蓬皮杜艺术中心，贾科梅蒂(Giacometti)画展的标题)
（6）绘画如梦 *(2005年罗浮宫和蓬皮杜艺术中心联展的标题)
（7）从历史的深度（表现）到当代的平面性

这门课同样也是我们和南京东南大学学生的第一次接触。因此，我们必须确保他们能很好地理解我们的想法。

这次讲座的第一幅图，是2009年9月李翔（建筑师，也是巴黎玛拉盖国立高等建筑学院的毕业生）在上海一间酒吧用马克笔在餐巾纸上画出的一副很小的画（图16）。他想将中国表现手法归为一个符号，而将欧洲和西方的表现手法归为一个小小的透视图，以此来展现两者之间的根本差异。实际上，我们还注意到，他本能地选择了一顶高高的西式礼帽来反映中国的表现手法，而用一顶中式帽子来反映西方的表现手法。这样一来，他诙谐地发现了中欧文化交流的复杂性，而且触及了本次教学活动的中心理念。

在这次讲课中，我们很多时候

2.3 Contents of the lectures

In this chapter we present the lectures as they were planned, following a plan with two sub parts. Then we'll go through the way the contents have been explained, in accordance with the theoretical classes of Nanjing Southeast University's workshop.

2.3.1 Lecture 1: A point in space

This first class was significant. It presented the spirit and the theoretical basis essential to the entire workshop, refering to the subject of the Western two-dimensional representation. At first we obviously went through the invention of the central perspective of the Italian Renaissance, but also prepared the students to have a better understanding of the specific evolutions which took place in that area, and to pay closer attention to today's ways of representation.

Plan of the conference

1st part
(1) First key points
(2) From Giotto to Matisse
(3) The First Renaissance
(4) The Second Renaissance

2nd Part
(5) *The work of drawing* (name of an exhibition of Giacometti's drawings at the Centre Pompidou in 2001)
(6) *The drawing dream* (name of a double exhibition at the Louvre and Centre Pompidou, 2005)
(7) From historical depths to the contemporary plane

This lecture was also the first contact we had with the Nanjing's school students, therefore we had to make sure they would clearly understand our ideas.
In addition, the first drawing introduced in this lecture was a small drawing (Figure 16), done with a marker pen on a napkin in a bar in Shanghai in September 2009 by Jean Li (Xiang Li), architect and former student at the ENSAPM. He wanted to show the main difference between the Chinese way of representation, by the use of a sign, and the European and Western way of representation, by the use of perspective in a small drawing. In fact, we can note that when he was doing his Chinese presentation he was wearing a top hat; and during his occidental and European presentation he was wearing a Chinese hat. By doing so he cleverly underlined the complexity of the Chinese and European cultural exchanges, and therefore touched on the central idea of our workshop.
Several times during this lecture we've shown some examples of Chinese paintings (Figure 17), and the peculiar painting of Giuseppe Castiglione (Figure 18) and Piero della Francesca(Figure 19).

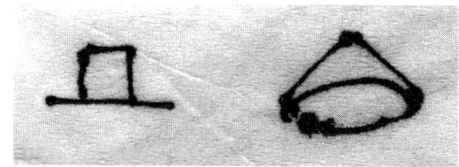

图16 餐巾纸上画出的毡帽

分析了中国绘画的几个例子（图17）以及郎世宁（Giuseppe Castiglione）（图18）、弗兰切斯卡（Piero della Francesca）（图19）稍显特殊的绘画。

当然，标题的选用直接指向了中央透视法的灭点（图20,图21），同样也直接指向绘画过程开始的时候在工具（比如说铅笔的笔尖）和载体（传统的纸张）之间产生的关联点。这首先能让我们自觉地理解英语称为"drawing"和"design"的内涵。当然了，我们同样也可以用计算机软件以及根据一个点确定的线条（即我们所说的坐标来思考）。

图18 郎世宁Giuseppe Castiglione(1688—1766)的作品

2.3.2 课程二：围绕朱利欧·保利尼(Giulio Paolini)的作品，从历史深度(表现)到当代的平面性(课程一的最后一个部分)

第二课，本来并不是设想按照这种形式进行的，但它却是理论和实践之间成功过渡的首要因素。前一天，包括昨天下午我们的谈话，昨天晚上我们和整个教学团队的交流，让我们产生了保证每天一节理论课的想法（原本预定是两天一节），同时缩短讲座的时间。这第二课很短（40分钟），直接将学生们带入2002年阿姆斯特丹当代艺术博物馆临时展览的现场。

介绍的第一幅作品朱利欧·保利尼(Giulio Paolini)的《正宗歌剧》。这幅画是在2001年完成的，展览期间就由博物馆购买收藏。我们总共用了约十五幅作品介绍了二十世纪六七十年代，并且通过这些通常为白色的、极少主义的作品，我们可以观察前一天刚研究过某些定义和分析过的社会背景的逐步走向。这种形式的课程，同样让我们能够在下午的指导性学习中，提出这些问题，并顺势将这些问题展开。

在第二次教学之后，我们整个教学小组一起决定，整个一周都保持这样的交替的讲座，一种是大型授课，基本按原先制定的授课内容，另一种是比较简

图17 中国画家仇英Qiu Ying（1498—1552）的作品

图19 弗兰切斯卡 Piero della Francesca (1416—1492) 的作品

The title of this first class is linked to the vanishing point of the central perspective (Figure 20,21). But also to the point where the tool (the point of a pencil for example) and the support (the traditional piece of paper)make contact and the actual work of drawing begins. We can then introduce the subject of the drawing and the design. Of course, we can also apply this to softwares in which the lined route is determined by a point defined by particular coordinates.

2.3.2 Lecture 2: An exhibition of Giulio Paolini's works, from the historical depths to the contemporary times (last part the first of lecture)
This second lecture, which at first wasn't meant to be this way, ended up being the crucial point for linking theoretical knowledge to practical knowledge. The previous day, after our afternoon talks and evening meeting with all the teaching staff, we came to the conclusion that each day we would have one theoretical class (instead of one every two days), in order to reduce the length of each intervention. This second lecture was short (40 min maybe), and brought the students all the way to a temporary exhibition at the Contemporary Art Museum of Amsterdam in 2002.

The first work presented was called *Opera Autentica* of Giulio Paolini. It was made by the artist in 2001 and then its acquisition by the museum was at the basis of this exhibition. In total, we used about fifteen works introducing the 60s and 70s, and thanks to these mainly white and minimalistic works, we could observe the progressive departure from the traditional framework, which we had defined and analysed in the lecture of the previous day. Structuring the lecture this way allowed to introduce some points that we would rephrase and develop during the afternoon tutorial.

After this second lecture, we decided that

图20 皮萨奈罗 Pisanello (1395—1455) 的作品

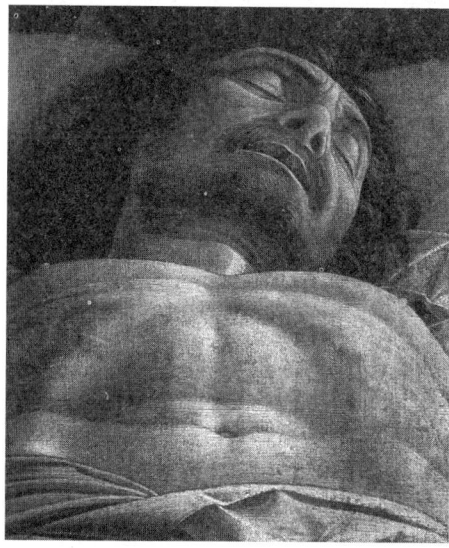

图21 曼坦那 Mantegna (1431—1506) 的作品

单、相对即兴的授课，也取决于学生们提出的问题，一方面是尝试对此作出回答，另一方面则是展开在学生自己练习过程中提出的各种课题。

2.3.3 课程三：只是思想（转化为作品）

第三课（讲座）的原则就是，对艺术作品的概念以及20世纪发生的根本性变革进行思考，同时又全面考虑到自19世纪以来预示这些变化的重大事件，尤其是第一批博物馆的诞生。

讲座提纲：只是思想（转化为作品）

第一部分：杰出作品和博物馆
（1）杰出作品一览
（2）分散的秩序、继承和遗产
第二部分：艺术和技术，开放的边界
（3）追求完美，劫持和消亡
（4）复制-粘贴或是文化和沟通

对两个世纪以来西方艺术思想的回顾（这一浏览非常概要）能够帮助学生进行自我定位，或者至少是能够参与辩论。如今这种辩论以历史性、现代性和当代性等概念为理论视角；以传统技术方法、创新性的新科技方法、或者越来越多的这两者的混合杂交，作为实践角度。这就是要为一个辩论提供某些元素，并发展一种批判的思想。这种思想是关于当今艺术现状，以及其在当今全球化的形势下，与社会和政治相关联的演化。

这种方法在诺曼底国立高等建筑学院需要通过20节课来实现，每节课一小时。最近，意大利的卡利亚里（Caglirari）建筑学院也采用了类似的形式。

相反，对于南京的这个教学活动，我们需要以非常概要、非常简单的形式对它重新进行全面考虑。当然，这同样需要和前面的讲课串联起来。

第一部分以罗浮宫的创建，尤其是1803年古希腊古罗马展厅重新开张作为开场。在大革命时期过后（1793年），这座博物馆世俗的百科全书式的功能直至今天都充满了勃勃生机……（图22）

这节课的第二部分，主要是聚焦着艺术、科学以及促成最初世界博览会的重大科技发明之间的互动关系。

大量的新信息，有点淹没了学生。然而，通过接收到的这些信息（所有重要艺术家的名字在上午的课上给了学生），通过课上拍的图片，以及通过因特网上的某些根据个人兴趣所进行的搜索，一部分学生已经逐渐知道了如何去着手相关的课题研究。

图22 画作内外的人物（罗浮宫）

for the rest of the week we would alternate between long and important lectures — more or less the way they were suppose to be — and some lighter classes, more or less improvised, based on the students' questions, in order to try and give them answers, but also to develop problems that appeared in their own works.

2.3.3 Lecture 3: Nothing but the Idea! (transformation at work)

The principle of this third lecture was to conduct a reflection on the notion of the work of art and radical changes that took place during the 20th century, but still kept in mind the events that happened before these changes, ever since the 19th century, including the creation of the firsts museums.

Lecture plan: Nothing but the Idea! (transformation at work)

1st part: The Masterpieces and the museum
(1) Overview of the masterpieces
(2) Random order, legacy and successions
2nd part: Art and techniques, open borders
(3) Quest for the absolute, abductions and disappearances
(4) Copy and paste, or Culture and communication

This quick overview covering two centuries of Western artistic thought, was intended to help the students place themselves in the centre of the debate concerning, today, from a theoretical point of view notions of historicity, modernity, contemporaneity and from a practical point of view of some traditional technical approaches as well as innovating technological approaches and a combination of those two processes. The aim here is to give some elements for a debate, and develop critical thinking concerning current artistic issues and the evolution of their social and political implication in globalisation.

This approach was designed for a schedule of 20 classes of one hour each for the ENSA of Normandy. The approach was used lately in the Faculty of Architecture of Caglirari in Italy, following more or less the same schedule. However, for this Nanjing workshop we had to think of a much shorter and simplified version. It also needed to fit in amongst previous lectures.

The first part starts with the creation of the Louvre Museum and mainly the reopening of the antiquities galleries in 1803. After the French Revolution period (1793), the Louvre was given a nonclerical and encyclopaedic purpose, which it still has today…(Figure 22)
The second part of the lesson was focused on the relations that emerged between art, science and the great inventions of a century which has witnessed the first universal exhibitions.

This great number of new information could destabilise the students. However, with this new information (all the important names were given at the beginning of the lecture), based on photos taken during class, and then on some personal research on the internet according to the students' personal interests, quite a few of them managed to come up with some quite interesting topics of research.

2.3.4 课程四：只是思想（转化为作品）（续课程三）

这节课最初与前一天所作的展示有着紧密联系，同时也注意到其独立性，并且考虑到了学生习作的初步结果。因此，在探究了19世纪之后，我们步入了20世纪（的艺术）。

以浏览我2002年为巴黎玛拉盖国立高等建筑学院课程所拍摄的华盛顿当代艺术博物馆作为开场。我精心拍摄了一系列作品（主要是20世纪60年代至今的画作），大多是正面照，但也有一些侧面照片，来展示绘画中视角的变化对动态视角的传达。因此，我们继续前面一课与阿姆斯特丹展览（朱利欧·保利尼 (Giulio Paolini)的作品）相关的内容，但这次展览是以传统手法拍摄的。

随后，我们集中关注一个以1906年塞尚（Cézanne）去世以及1907年他的作品回顾展作为起点的时期。我们研究了人物形象、形式和背景的关系演变，主要以马蒂斯和毕加索为例，随后是形象的逐步消失，以及抽象造型的诞生，正如我们可以在蒙德里安和马列维奇（Malevitch）的作品中看到的那样。我们主要研究从1906—1930年这个具有决定性的意义的时期，当时巴黎在这个时期曾是首屈一指的国际文化的聚集地。最后，我们还向马塞尔·杜尚（Marcel Duchamp）致了敬……

我们同样还对如今众多的复制手段颇感兴趣，既包括机械复制（沃尔特·本雅明 Walter Benjamin），同样也包括工艺复制。

在这个过程中，我们还给出了一个由小图片组成的全景式的一组图片形成的一个单一文件。这个文件能够让我们观察到横跨150多年的时期内整个一系列作品之间可能的联系。

最后，作为总结，我们在这个上午首次向中国学生展示了法国学生作业的几个案例。这些案例针对的是不同的练习，但具有可比性。

2.3.5 课程五：绘画的时间

第三个讲座，遵循了和之前两个讲座相同的逻辑，但这一次，聚焦的是我的个人作品，尤其是我的绘画作品（图22—图24）。我们也把它们和先前讲座里研究过的其他的艺术家、其他作品以及我的教学建议联系起来。

讲座提纲：绘画时光

第一部分
（1）平面的深度
（2）欧洲属性
（3）向现代主义的学习
（4）向传统的学习

第二部分
（5）分割的方形或双重的方形
（6）多折画
（7）绘画的追索
（8）画家和建筑师

这门课按照我们之前计划的那样进行，因为其性质稍微有些不同，是根据我在2009年分别在大连和上海所

2.3.4 Lecture 4 : Nothing but the Idea or (transformation at work) (End of Lecture 3)

The class was initially meant to continue on the theme of the class of the previous day, Which was then revised to be more autonomous and include the students' concrete results. Therefore after exploring the 19th century we stepped straight into the 20th.

We started with a virtual walk through the Washington Museum of contemporary art through some photos taken by me in 2002 for my classes at ENSA Paris-Malaquais. I had taken some pictures of works (mainly from the 1960s till today), from the front of course, but also from the side, to show how the change of perception on the painting can convey a dynamic point of view. Thus, we continued on the subject of the previous lecture, about the Amsterdam exhibition (Giulio Paolini's works) — but the pictures from that exhibition were taken in a more traditional way.

Then we focused on a period starting after the death of Cezanne in 1906 and his retrospective in 1907. We studied the links between the figure, the shape and the background, mainly with Matisse and Picasso, then the progressive extinction of the figure, and the birth of abstraction, that we can notice in Mondrian and Malevitch's works. We mainly focused on the years 1906 to 1930, when Paris was a key place on the international scene. Finally we had a look at some of Marcel Duchamp's works.

We also played closed attention to many modern ways of mechanical (Walter Benjamin) but also technological duplication.

We went through this with a series of photos presented as vignettes on a single document, which allowed us to observe a few common points within a series of works on a period of time of more than 150 years.

Finally, that same morning, we had the opportunity to show to the Chinese students the work of some French students on similar exercises to the ones they had done.

2.3.5 Leture 5: The time of painting

This lecture also followed the logic of the two previous lectures, but in this one I focused on my personal work and paintings (Figure22-24). We connected these to some artists and works studied in previous lessons, but also with my personal pedagogical propositions.

Plan of the conference: The time of painting

1st Part
(1) The depth of the surface
(2) A European legacy
(3) Modern learning
(4) Classic learning

2nd Part
(5) Divided square or double square
(6) Polyptychs
(7) The claims of drawing
(8) The painter and the architect

This class was presented according to

作的有关我本人作品创作的讲座改编而成的。这个讲座介绍的是我艺术作品主题的发展变化,作为参考学生能再次看到那些重要的作品。这些作品在前面的课上,最初是放在历史背景中介绍的,而这一次是以更加主观的方式,以组合、对比、致敬或是它对我个人的实践中启发等形式来展现介绍。

此外,周末前一天的最后一节课也让学生们能够了解到我本人的艺术定位,这有可能会使得之前的课程更加清晰易懂,一是明白其内容所具有的"历史客观性",二是明白选择这些内容作为此次教学活动的目的所在。

2.4 学生们的反应和与他们的对话

第一节理论课学生们非常用心地听了,随后,其中的许多人感到了困惑。这种情形在第一天的下午一直持续着。因首先是试点小组的学生,随后是其他小组的学生不停的来征寻求更多的解释,甚至来问要求他们在第一次练习中找出的答案。

然而,我们会感到,总体上,很多学生都能在研究探索过程中,对理论课的内容以及图片作出比较积极和正确的回应。

从学生们对最初两节理论课的感性认识,到第二天的习作辅导课,我们想了一个称为"三词实验"的主意,就是让学生很快的,快速思考之后,写下三个能充分地表达课上给他们印象最深的图片、信息或者想法的

图22 平衡,作者画稿,2006

图23 构成,作者画稿,2006

the plan we made as it was a bit different from the other classes, being based on a lecture about my work that I held in 2009, in Dalian and Shanghai. In addition to presenting some of my works of art, this lecture allowed students to go through some important works, initially presented in their historical context, but now seen from a more subjective angle, in association and confrontation, with the inspiration and developments that they had brought to some of my works.

In addition, this last class, before the weekend, informed the students about my personal artistic point of view, and may have brought some more precisions to the previous classes, regarding their "historical objectivity" as much as our pedagogical positioning for this workshop.

图24 作者在巴黎的展览，1986

2.4 Students' reactions and dialogues

The first theoretical class was followed very attentively by the students, but there was a moment of distraction for many of them. This lasted all afternoon, as some students from the pilot group, and then students from other groups, kept asking for additional explanations, and sometimes for the answers that they were supposed to look for and find out by themselves in the first exercise.

However, we could notice that generally many students comprehended well and quickly what was said during the theoretical classes and shown in the pictures, and went on to achieve appropriate research.

This lead us to decide that on the tutorial of the second day we would use the students' sensory reactions during the first two theoretical classes, through what we called "the three words experiment", consisting of simply writing down three words that they felt were the most appropriate to describe their strongest feelings and reactions to the images, information or ideas that were presented during the previous lessons. Some of them only came up with one word, but it was sufficient.

Surprisingly, I easily created a genuine interaction with the students, sometimes just by using English, but most of the time through Wang Ying's translation from French to Chinese — he would always rapidly find the right words in each of the two languages.

Many of them were at ease with the exercises and accepted to go through with their project freely, as encouraged by Mr Zeng.

Then, the first positive results encouraged many others, and from the pilot group we were able to do vey good work with the rest of the year.

Then everything went very fast, sometimes too fast, with a rapidity in execution and very diverse means. Some of the projects showed evidence of very good technical realisation. There were some installations, but because of the lack of time, they were usually not carried out quite as well. There was humour and pertinence in some answers, and many ideas. Several times fire was used — that was something I had never seen with European students.

词语。有时候一些学生只有一个词,但也足够了。

这样一来,我可以很容易地和学生们建立起真正的交流。这有些出乎我的意料。这种交流有时可以直接用英语,但大部分的时间通过王盈先生将法语译成汉语——他总是很快就能从这两种语言中找到合适的词。

在曾先生的鼓励下,他们中的大部分人自在地完成了这些练习,并且以一定的自由和自主参与了习作。

随后,第一次习作有很好的结果。这鼓舞了很多其他人,从试点小组开始,我们真正能够和整个一届的所有学生一起工作了。

然后,随着学生们迅速的制作速度和真正多样化的手段,一切进展都很快(有时可以说是太快了)。有很多作品,其中有些在工艺制作方面是非常成功的。但确实有些装置作品,因为时间不允许,最终没能完全成功。有些作品非常幽默,非常恰当,充满着想法。还有很多学生都用到了火——这是我在欧洲学生那里从没有见到过的。

2.5 与中方教师的对话以及作出可能的调整

对话是持久的,经常性的也是逐步进行的。如今我们仍然在朝着这个方向努力。

然而,我们之前提及过,东南大学的重要期望之一,就是在教学上带来一定的方法,并能够与现存的教学方式很好地协调一致。我们走到一起教学,就是以此为目标,来达到一种一致性:

·(不会讲汉语的)欧洲教师带来的信息建议与中国学生可能及真正理解的内容之间的一致性。

·我的中国同事根据几个与学生更直接的范例所传达出的建议的一致性。

·一个为期10天的教学活动和东南大学一年的中长期观察的教学之间的一致性。

2.5 The dialogue with the Chinese teachers and eventual adjustments

The dialogue was continuous, regular and progressive. It still continues today just in the same way.

 As we have already seen, the main goal is to bring to the Southeast University some methodological advice and good coordination between the existing courses. We were therefore gathered in order to achieve these goals:
· coherence between the information proposed by a European teacher (who doesn't speak Chinese) and a potential and real assimilation by the students.
· coherence between ideas passed on by my Chinese colleagues from some direct examples from students
· coherence between the ten-day workshop and an annual course, with an observation of the short and long term by Southeast University.

2.6 法文原文

1 Enseignement de l'art dans les écoles d'architecture en France

1.1 Art et architecture
1.1.1 Quelques repères historiques européens

Historiquement dans la culture occidentale, à partir du début du XVème siècle en Italie, à Florence précisément, un mouvement, qu'on appellera plus tard la Renaissance (Burckhardt, 1860, *La civilisation de la Renaissance en Italie*), va rassembler des architectes, des sculpteurs, des peintres qui vont révolutionner les codes de la représentation. La pierre angulaire de cette pensée nouvelle était une invention théorique, la perspective, fondée sur l'usage du dessin, *Arti del designo*, lequel allait égaler en importance la rhétorique et la poésie, fondées elles sur l'usage du verbe.

La terminologie *Beaux-Arts* qui nous est familière viendra plus tard au XIXème siècle. Il y aura d'ailleurs une architecture de style *Beaux-Arts*, néoclassique et d'influence française, qui s'exportera dans le monde entier et notamment aux USA dans des villes comme NYC ou Chicago. Le bâtiment de l'Université du Sud-est répond lui-même à certains critères formels de cette époque.

Paradoxalement, pendant la même période, en Europe et à Paris, particulièrement entre les années 1905 et 1930, la rupture de la Modernité est particulièrement féconde. Ce nouvel affranchissement vis-à-vis des codes de représentation, issus cette fois de la Renaissance, entraîne les artistes vers de nombreuses expérimentations et découvertes. L'utilisation des papiers collés, l'invention du ready-made, le choix de l'abstraction sont autant de libertés nouvelles avec lesquelles l'homme interroge le monde et participe, par des voies très différentes, de cette sortie progressive du plan du tableau et de son cadre traditionnel.

Ce qu'on appellera communément plus tard les *arts plastiques* favoriseront des collaborations intenses entre peintres, sculpteurs et architectes. Le Bauhaus, De Stijl, le Constructivisme en sont des exemples illustres qui marqueront les esprits. Au moment où nous rédigeons cette publication, une remarquable exposition au Centre Pompidou présente justement l'œuvre de Piet Mondrian et le mouvement *De Stijl*.

Ensuite, pendant la seconde guerre mondiale, beaucoup de grands artistes européens émigreront en Amérique du Nord et plus particulièrement à NYC. L'art de la seconde moitié du XXème siècle est donc fondamentalement marqué par l'influence américaine. Toutefois, les échanges entre L'Amérique du nord et la vieille Europe resteront très actifs. En 1977, le Centre Georges Pompidou fut inauguré avec une exposition emblématique sous le titre *Paris-New-York*. C'est avec un titre tout aussi révélateur, *The American Century - Art&culture 1950-2000*, que le Whitney Museum décida de clore le millénaire en Septembre 2000 par une importante rétrospective sur l'art américain.

Aujourd'hui en 2011, c'est évidemment avec un regard immergé dans la mondialisation qu'il nous faut observer les changements en cours, les transformations en devenir pour et dans lesquelles la Chine jouera certainement un rôle très important.

1.1.2 Quelques repères institutionnels français

En 1870, il y avait en France sous la Troisième République, un Ministère de l'Instruction publique, des Cultes et des Beaux-arts. En 1881, Jules Ferry fut le nouveau Ministre l'Instruction publique et des Beaux-arts, et effectivement, c'est bien pendant ce XIXème siècle que la notion d'œuvre d'art va considérablement évoluer et cessera de répondre de manière prépondérante à des aspirations religieuses.

Plus tard en 1959 et à nouveau en 1962, André Malraux fut nommé pour la première fois par le Général de Gaulle, Ministre d'État, chargé des Affaires culturelles, donnant ainsi à la culture une autonomie nouvelle.

Cette particularité française fut renforcée par François Mitterrand en 1981, notamment par l'importance des crédits, avec la nomination de Jack Lang, qui fut à son tour pour une première fois, Ministre de la Culture.

Actuellement, nous héritons de cette situation avec deux ministères distincts, le Ministère de l'Education Nationale, mis en place dès 1932, et le Ministère de la Culture et de la Communication, régulièrement reconduit sous cette dénomination depuis 1997.

L'évolution est significative mais sur le site historique de la rue Bonaparte à Paris, il y a toujours actuellement l'Ecole Nationale Supérieure des Beaux-Arts. Crée en 1819 sous forme d'Ecole royale des Beaux-arts, puis transformée en Ecole impériale en 1883, elle abrita en son sein l'enseignement de l'architecture conjointement avec celui de la peinture, de la sculpture et de la gravure jusqu'en 1968.

En effet, les évènements de 1968 marqueront une rupture politique et épistémologique qui sera significative en France. Elle entraînera à l'intérieur l'ENSBA un éclatement disciplinaire et par la suite l'enseignement de l'architecture sera dispensé sur différents sites par 8 unités pédagogiques d'architecture (UPA) très différentes les unes des autres

et réparties sur tout le territoire français. C'est André Malraux qui sera d'ailleurs à l'origine de cette décision pour répondre au rejet de l'académisme et du style *Beaux-Arts* qui sera très violent à cette époque.
Pendant quelques années, l'enseignement de l'architecture dépendra du Ministère de l'Equipement avant de revenir rapidement à la Culture.
D'une manière générale, cette organisation perdure actuellement. Toutefois, les unités pédagogiques sont devenues entre-temps des écoles nationales puis des écoles nationales supérieures d'architecture.
La plus récente d'entre elles, l'Ecole Nationale Supérieure d'Architecture Paris-Malaquais, crée sur le site historique en 2001 par un collectif d'enseignants, tente de remettre en place aujourd'hui des ébauches de partenariat avec l'Ecole Nationale Supérieure des Beaux-Arts. Cependant, la séparation reste encore très forte.

1.1.3 Le champ ATR (Arts et techniques de la représentation)

A l'intérieur de chaque école nous sommes passés après 1968 d'un enseignement d'atelier dit *vertical* à des enseignements interdisciplinaires dits horizontaux et rassemblés autour du projet. Il y a au total 6 champs disciplinaires qui font tous l'objet de recrutements spécifiques dans le cadre de concours nationaux :

- Théories et pratiques de la conception architecturale
- Histoire et cultures architecturales
- Sciences et techniques pour l'architecture
- Villes et territoires
- Arts et techniques de la représentation
- Sciences humaines et sociales

Toujours selon la dénomination officielle dans les écoles d'architecture en France, le champ ATR, (Art et Techniques de Représentation), est lui-même divisé en deux sous-champs :
- Techniques de représentation
- Arts plastiques et visuels

Cette séparation est parfois un peu dommageable car elle peut ponctuellement entraîner un écart trop grand entre l'outil et l'idée. Cela conduit parfois les étudiants à ne pas toujours pouvoir exprimer leur pensée ou au contraire à être tributaire d'un savoir faire technique qui conditionne prématurément leurs projets.

Quoi qu'il en soit, les apports théoriques sont généralement mieux assimilés en alternance assez fine avec des expériences pratiques et personnelles. Les étudiants peuvent alors repérer plus facilement les acquisitions nécessaires et pertinentes au regard de leurs études et se positionner progressivement face aux métiers de l'architecture selon des choix à la fois conceptuels et techniques. Il y a d'ailleurs une vraie demande étudiante allant dans ce sens à laquelle il faut répondre pour leur permettre de développer aux mieux leurs compétences futures face à l'évolution des métiers de l'architecture et de la ville.

1.2 Méthodologie générale et approche personnelle
1.2.1 Peintre et architecte, une double formation

De l'âge de 11 à 17 ans j'étudie le dessin dans une école d'art et je me passionne pour la peinture dans les musées et à travers les livres. A 17 ans, je découvre l'œuvre de Mondrian. L'année suivante, en 1970, je commence mes études d'architecture à Paris. Ce sera l'occasion d'observer certains rapports entre l'art et l'architecture et aujourd'hui encore je poursuis cette recherche à travers mes activités professionnelles.
En 1976, lorsque je passe mon diplôme d'architecte (titre *: L'architecture et le rêve,* sous-titre *: Les rapports entre peinture et architecture*), je demande à André Fermigier* d'être membre de mon jury. Il accepte ma sollicitation en me précisant toutefois : "La peinture et l'architecture sont de faux amis" et d'ajouter, "vous verrez, ce ne sera pas facile". Effectivement, il me faudra de nombreuses années pour ne plus aborder l'espace pictural comme un architecte mais bien comme un peintre. Ce n'est pas le même espace. Ce n'est fondamentalement pas la même discipline.

*André Fermigier (1923-1988) fut historien, critique d'art, professeur. Il fut l'auteur de plusieurs ouvrages notamment sur des peintres importants, Picasso, Bonnard, Courbet etc. Il écrivit aussi des chroniques d'architecture et d'urbanisme des années 1960 au début des années 1980, rassemblées dans un livre, *La Bataille de Paris*, Gallimard, 1992.

Aussi, de 1976 à 1994, je décide de consacrer tout mon temps à la peinture et d'exposer en Europe. Par contre et c'est aussi un point important, toutes mes expositions sont une mise en relation de la peinture avec l'espace architectural. C'est seulement à partir de 1995 que je choisis d'enseigner dans plusieurs écoles d'architecture. Depuis 2002, à travers des conférences et des workshops, j'affine ma démarche artistique et mes approches pédagogiques en reliant de plus en plus mes différentes activités professionnelles.
Nous l'avons évoqué précédemment, la Modernité, particulièrement pendant la période de l'entre deux-guerres, a certainement renouvelé quelques passages évidents entre les deux disciplines mais on doit clairement, aujourd'hui au XXIème siècle, repositionner le débat entre art et architecture et aborder l'évolution de

ce que nous appelons aujourd'hui encore "l'art contemporain". On peut d'ailleurs raisonnablement s'interroger sur la définition historique et la pérennité de cette expression "art contemporain" qui domine la scène internationale depuis les années 80.
On notera pour préciser les termes et enrichir le débat que Catherine Millet*, critique d'art (*Artpress*) et écrivain: "… Car ce n'est pas l'art qui tend à une fonctionnalité dans le design, la photo documentaire, la mode, etc., ce sont ces activités qui cherchent à acquérir le statut de l'art . Aussi, tenons-nous en à cette définition : l'art contemporain est un pôle d'attraction. Rien de plus, rien de moins. "
*Catherine Millet, *L'art contemporain, histoire et géographie*, Flammarion, 2006.
Enfin on remarquera, lors d'un colloque "art et architecture" organisé à Lyon en novembre 2010, que Martine Bouchier*, architecte, docteur HDR en esthétique et professeur à l'ENSAPVS, déclarait que "l'architecte et l'artiste n'ont pas vocation à collaborer". Bien sûr, au-delà d'une provocation salutaire sur la transversalité, Martine Bouchier pointe avec précision et justesse des confusions trop souvent entretenues et dégage deux notions distinctes en glissant sémantiquement de l'œuvre d'art à l'objet culturel en fonction de leurs relations réciproques aux institutions.
*Martine Bouchier, *L'art n'est pas l'architecture*, Archibooks, 2006.

Aujourd'hui, de l'objet culturel à l'objet de culte, la distance est parfois de plus en plus mince.

1.2.2 La transversalité pour enseigner

Au regard de ma formation d'architecte, de mon parcours de peintre et de mes expériences actuelles liées aux arts vivants, je rapproche progressivement ma pratique d'artiste et mon travail pédagogique en développant de manière croisée des questionnements théoriques qui abordent la représentation tant dans l'espace plan que dans l'espace tridimensionnel.
Aussi, pour résumer mon opinion en ce qui concerne la place qui est accordée aux enseignements des disciplines artistiques dans les écoles d'architecture, il me semble, d'une manière générale, que l'on peut les regrouper dans deux directions principales:
- des enseignements artistiques autonomes dispensés sous forme de cours spécifiques et qui n'ont pas d'implication directe sur un projet architectural mais qui participent d'abord de l'acquisition d'outils et de connaissances liées aux techniques de représentation et ensuite d'une culture générale, les deux étant nécessaires à la formation des architectes.
- des enseignements transdisciplinaires proposant un détour artistique mais ensuite impliqués dans l'élaboration d'un projet architectural. Ces enseignements ne sont pas forcément immédiatement opérationnels mais posent la question du retour au projet, de l'appropriation de l'expérience vécue et de savoirs venant enrichir son processus créatif.
Par essence même, l'interdisciplinarité et donc la transdisciplinarité impliquent l'existence de champs disciplinaires et le développement d'une pensée qui leur est spécifique. Pour être véritablement bénéfique quant à l'enrichissement que cela apporte pour la recherche de l'étudiant, le voyage doit être effectué jusqu'à son terme avec un retour complet dans la discipline de départ où on retrouve les règles qui lui sont attenantes. C'est un point très important car trop d'étudiants, séduits par le changement qu'entraîne la transdisciplinarité, et je l'observe dans mes enseignements, peuvent parfois être tentés par des comportements, par des formes dont ils n'ont pas eu le temps, pendant ce voyage, de comprendre et d'assimiler l'origine. Il ne s'agit pas de voler la liberté découverte dans le champ voisin mais bien de s'approprier en architecture celle qui lui appartient en propre et lui donne sa véritable dimension.

1.3 Mes objectifs comme enseignant ATR dans les écoles d'architecture

Aujourd'hui, de manière simultanée, se côtoient dans l'art des approches traditionnelles et des approches technologiques innovantes que l'on retrouve dans bien des domaines de la vie et à commencer par celui qui nous concerne tous, la reproduction des espèces.
De fait, le rapprochement entre création et procréation passe par de multiples possibilités de transplantation ou de duplication partielle du corps humain et déplace le questionnement de l'origine, de la transgression et des représentations.
C'est donc à partir de ces remarques que je propose aux étudiants quelques objectifs clairs:
- Comprendre une appartenance européenne.
Mes enseignements relèvent de quelques fondamentaux du champ ATR, ils sont fortement ancrés dans l'histoire de l'art occidental et complètent une vision ouverte sur le monde par la fenêtre de nos ordinateurs via l'écran plat.
Ils se regroupent sous un titre générique: Présentation, représentation, présence.
1. Un point dans l'espace, (atelier dessin, licence)
2. De la profondeur historique au plan contemporain, (ateliers théoriques et pratiques, licence)
- Travailler à l'échelle du vivant.
Mes enseignements traversent la réalité de plusieurs champs artistiques. Le corps

humain en devient le véhicule privilégié de la pensée pour appréhender l'espace d'un édifice en expérimentant ce que Walter Benjamin* nomme «la réception tactile» qui, autant que la vue, fonde notre réception de l'architecture.
*Walter Benjamin (1892-1940), *L'œuvre d'Art à l'époque de sa reproductibilité technique*, 1939 (première version en 1936).
3. Le corps à l'édifice, (enseignements thématiques et studio de projet, master)
4. La transformation à l'œuvre, (enseignements questions théoriques et travaux dirigés, master)

C'est une évidence, les approches technologiques innovantes sont incontournables aujourd'hui dans les écoles d'architecture. Mais pour autant, innover dans cet environnement pédagogique, c'est aussi être capable de mettre en place un clavier suffisamment large de connaissances théoriques et pratiques pour prendre en charge, in fine, des rapprochements essentiels, ou qui peuvent le devenir, entre apports historiques, modernes et contemporains.

C'est sur cette "trilogie" que se définissent mes enseignements qui se modulent selon les établissements et tentent d'aborder certaines contradictions apparentes ou réelles, voir certains paradoxes, "les préjugés de demain" selon Marcel Proust*, afin d'aider les étudiants à mieux se positionner face aux enjeux actuels, dans un premier temps, à l'intérieur des écoles d'architecture et bien sûr ensuite à l'extérieur face l'évolution des métiers de cette discipline en pleine mutation.
*Marcel Proust (1871-1922), "Les paradoxes d'aujourd'hui sont les préjugés de demain". *Les plaisirs et les jours*, 1896.

1.4 Expériences singulières en fonction de deux écoles en France

1.4.1 Exemple d'enseignement ATR mis en place en 2008 à l'ENSA de Normandie

Un point dans l'espace

Objectifs
Cet enseignement fut défini pour des étudiants de 1ère année et à ce titre, il est déterminant puisqu'il met en place un fondement concernant des questions liées à la représentation à l'ENSA de Normandie. Il commence par un cours magistral, qui pose quelques enjeux essentiels de l'espace occidental, et c'est d'ailleurs avec ce cours que nous avons commencé le workshop de 2010 à l'Université du Sud-est.
L'objectif est d'assurer une première base théorique et un apprentissage pratique autour de plusieurs modes de représentation, notamment le dessin, en favorisant:
- Une confrontation entre l'échelle du réel et les échelles de la représentation.
- Une compréhension du passage entre espace tri et bidimensionnel et vice versa.
- Des acquisitions de connaissances historiques et théoriques sur l'art et les techniques de représentation.
- Une justesse d'observation et de restitution en adéquation avec les moyens mis en œuvre.
- Une liberté de l'imaginaire s'appuyant sur des outils favorisant une expression personnelle.
- Une exploration vers des champs nouveaux, disciplinaires, artistiques ou techniques, en privilégiant de plus en plus le corps humain lui-même comme véhicule d'investigation vers de nouveaux espaces.

Mode pédagogique
Dans le cadre des exemples présentés, le cours magistral est suivi d'une visite au Musée des Beaux-Arts de la ville de Rouen, et les étudiants sont invités, à partir d'un tableau librement choisi, d'en faire quelques dessins libres, puis une première maquette en volume pour la séance suivante. L'exercice dure en général 5 à 6 semaines et alterne des rendus tridimensionnels et bidimensionnels. Au fur et à mesure de leur avancée, les corrections sont d'abord collectives puis très individualisées. En fonction de leur démarche, chaque étudiant est encouragé à trouver progressivement un médium qui soit en adéquation avec ses intentions et les découvertes plastiques, recherchées ou accidentelles, qui apparaissent au fur et à mesure de l'évolution du travail. Un texte méthodologique, assez simple et accompagnant le projet, est demandé le jour du rendu.

Exemples
Les exemples choisis présentent 3 cas différents, le premier est celui d'un étudiant qui possède une très bonne technique sur laquelle il peut s'appuyer pour conceptualiser de manière radicale tout en le faisant de manière fine et subtile. Le second, au contraire, est celui d'un étudiant qui n'a reçu aucune formation particulière, et dont le dessin reste pauvre, mais qui par contre, a su compenser par une démarche intuitive, jouant directement sur la couleur et la matière, et intégrer de manière sensible les références qui lui ont été données. Enfin, le troisième exemple est celui d'un binôme, c'est-à-dire de deux étudiants, qui ont un niveau technique moyen mais qui ont su transcender leur réflexion de départ et oser une échelle de représentation liée à une performance de leur propre corps comme outil de représentation.

Exemples 1 à 3
Voir page 18 à 23

1.4.2 Exemple d'enseignement ATR mis en place en 2002 à l'ENSA Paris-Malaquais

De la profondeur historique au plan contemporain

Objectifs
A l'ENSA Paris-Malaquais, cet enseignement fut généralement donné en 3ème année de 2002 à 2006. Il fut modifié en 2008 pour l'ENSA de Normandie et récemment adapté pour le workshop de l'Université du Sud-est de Nanjing en 2010.
L'objectif est d'amener les étudiants, lorsqu'ils observent l'art moderne, et plus particulièrement l'art contemporain, à ne pas s'immobiliser dans l'actualité afin de développer une interrogation critique sur l'origine. Avoir la curiosité d'une chronologie étaye durablement une attitude prospective sur le futur.
A l'appui de quelques connaissances et repères iconologiques et chronologiques, cet enseignement, en utilisant d'ailleurs le flash-back dans le déroulement de sa présentation, propose aussi aux étudiants une lecture ou une relecture de l'art contemporain en le situant dans le champ historique d'une investigation élargie. Chacun des six cours aborde une thématique précise que l'on définie peu à peu au regard d'un ensemble d'œuvres, que ce soit celles d'une période, d'un mouvement ou de quelques artistes sur lesquels on concentre notre attention.

Mode pédagogique
Ces cours théoriques sont immédiatement suivis par la proposition d'un sujet d'exercice à partir duquel l'étudiant produit une mise en œuvre concrète (en 8 jours) et un texte bref (1 à 2 pages) à la fois réactif au cours et explicatif de la démarche suivie pour le choix de la réponse. Les énoncés sont directement liés à la thématique du jour et aux questions ou discussions qui ont suivi. Ils ont pour but de provoquer le regard critique des étudiants et de les entraîner à être plus conscients des enjeux lorsqu'ils travaillent dans ou avec l'espace à 2 ou 3 dimensions pour les aider ensuite à plus d'aisance dans les manipulations qu'ils réalisent dans leurs studios de projet.
La correction collective de l'exercice pratique est un moment très révélateur de la pluralité des réponses et apporte une plus large interprétation et une meilleure compréhension du cours précédent.

Exemples
Les exemples ont tous été choisis pour la qualité d'une démarche attentive à la pertinence d'une idée, et cohérente par son adéquation avec sa matérialisation. Ils répondent à des exercices distincts dont on retrouvera chaque intitulé en tête des textes brefs qui accompagnaient chaque rendu. Dans les premiers cas, les 4 exemples ont eu des intitulés légèrement différents et correspondent à des années universitaires toutes différentes. Dans les derniers cas, c'est au contraire à partir strictement du même intitulé qu'ont été réalisés ces travaux. A chaque fois, on observera une grande diversité des réponses proposées, diversité également très présente dans les travaux des étudiants de Nanjing.

Exemples 1 à 6
Voir page 26 à 37

2 Workshop à l'Ecole d'Architecture de l'Université Sud-Est

2.1 Préparation du workshop à Nanjing

C'est bien sûr à partir des observations et expériences précédemment décrites que nous avons établi des propositions en sachant que nous nous adressions à des étudiants chinois. Bien évidemment, Wang Ying avait un rôle d'intermédiaire tout à fait déterminant puisqu'il fut formé comme architecte une première fois à Pékin, une seconde fois à Paris. Au-delà de ses compétences de traducteur, sa double expérience d'ancien étudiant lui permet sans doute de traduire mais aussi d'interpréter en finesse les demandes explicites et implicites de nos interlocuteurs.
Dans un premier temps, nous avons essayé de bien comprendre quelles étaient les souhaits pour l'Ecole d'Architecture de l'Université du Sud-est. Nous avons délibérément beaucoup travaillé en amont pour établir une méthode d'élaboration progressive avec dialogue permanent afin de bien définir les attentes de nos partenaires chinois et leur mise en relation avec nos compétences et les possibilités matérielles mises à disposition en sachant que nous aurions à travailler avec un effectif important dans un temps relativement restreint.

2.2 Déroulé réel au jour le jour

Entre le l'organisation prévisionnelle et la réalité effective du workshop, nous avons procédé à quelques décalages et plus particulièrement en fractionnant des cours théoriques pour en améliorer la réception. Décrire le plus simplement possible de déroulé réel pour cette première publication paraît la meilleure manière d'en rendre compte en sachant qu'ultérieurement, l'enjeu souhaité par L'Université du Sud-est sera, progressivement et avec le recul du temps, d'en dégager principalement les aspects méthodologiques.

J0/ 8.23/
Arrivée à l'Université du Sud-est. Présentation du workshop et derniers ajustements des attendus et de nos propositions. Visite des bâtiments dont

la bibliothèque et découverte de quelques ouvrages précieux rapportés de France par Yang Tingbao.
La première réunion a montré que nous nous étions tous bien préparés à travailler ensemble et la question principale fut de choisir, soit de travailler avec un seul groupe de trente étudiants, soit de le faire avec l'ensemble de la promotion. Nous avons opté pour la solution la plus collective à partir d'un groupe pilote. L'ambition était plus grande, nécessitait de bons relais mais justement il nous a semblé que nous étions prêts pour ce choix.

J1/ 8.24 /
Le matin, la première conférence (2h30 environ) correspondait bien sûr très précisément à ce qui avait été annoncé et devait permettre à l'ensemble des étudiants une première base d'informations et de connaissances, nouvellement acquises ou révisées.
Pour le bon déroulement du workshop, il était nécessaire de poser cette première base théorique qui par ailleurs proposait quelques passerelles entre culture européenne et culture chinoise. Elle se terminait par la présentation de l'exercice n°1 à partir d'une citation de Paul Valéry*, écrivain, poète, philosophe et épistémologue français : "Ce qu'il y a de plus profond chez l'homme, c'est la peau".
* Paul Valéry (1871-1945), *L'Idée fixe ou Deux Hommes à la mer*, 1932.
Nous avons alors demandé aux étudiants de réfléchir à une proposition proche et adaptée à cet exercice :
"Rien n'est plus profond qu'une surface."
L'après-midi fut divisée en deux temps:
- une réunion avec le groupe que nous appellerons "pilote".
- une déambulation pour visiter ensuite chaque groupe et entamer un dialogue avec les étudiants et leurs professeurs respectifs. Ce fut aussi l'occasion de visiter leurs ateliers en leur présence. Le soir une réunion fut organisée avec les enseignants et à la fin de ces échanges, nous avons pris la décision de proposer le lendemain une intervention brève, environ 40 minutes, pour compléter et enrichir les propos de la veille.

J2/8.25/
Le matin, le cours supplémentaire, décidé la veille, fut effectivement assez bref et dura environ 40 mn. Il poursuivait le travail de préparation de la première conférence et ouvrait plus largement le champ d'investigation pour les étudiants. C'est sans doute cette intervention, plus ciblée et reprenant quelques œuvres d'artistes présentées la veille mais cette fois dans un autre contexte, celui d'une exposition, qui fut déterminante pour permettre aux premiers étudiants de se libérer.
L'après midi, je suis d'abord resté avec le groupe pilote et ensuite ce sont les étudiants des autres groupes qui le souhaitaient qui nous ont rejoints.
Le soir, une seconde réunion fut organisée avec les enseignants pendant laquelle j'ai montré des travaux d'étudiants français. Il s'agissait de définir les modalités de corrections du premier exercice. Par ailleurs, tous ensemble, nous avons confirmé plusieurs décisions et précisé quelques choix pédagogiques :
-la proposition de conférences régulières chaque jour mais plus brèves.
-la demande aux étudiants d'écrire trois mots clefs sur leurs projets pour les aider éventuellement à mieux cerner leurs propres souhaits et vérifier qu'ils s'y tiennent.
-l'intervention du directeur du département pour bien expliquer les enjeux et encourager clairement les étudiants à prendre plus de liberté. Cette intervention de Monsieur Zeng fut très importante.

J3/ 8.26/
Le matin, conférence n°2 (1h30 environ). Elle fut simplifiée par rapport à celle du programme initial et seule la première partie fut abordée.
L'après-midi, correction du premier exercice. L'ensemble des étudiants sont présents. Après un tour assez rapide, nous choisissons une trentaine de travaux selon quelques critères prédéfinis (pertinence du concept, qualité de la réalisation, cohérence, poésie...). Nous choisissons plutôt les meilleurs exemples mais pas uniquement. Ils sont ensuite corrigés collectivement devant tous à partir d'un échange avec chaque étudiant ou groupe d'étudiants concerné.
A la fin de cette correction, nous donnons le thème de l'exercice n°2 faisant suite au cours théorique du matin qui sera à rendre pour le surlendemain. Il s'agit cette fois d'une citation de Xavier Fabre*, architecte et enseignant à l'ENSAPM : "toute architecture est un cadrage du mouvement".
*Xavier Fabre, *Repères*, n°18, publié par le Centre de Développement Chorégraphique et la Biennale de Danse du Val de Marne, 2006

J4/8. 27/
Le matin, suite de la conférence n°2 (40 mn environ), réellement modifiée et simplifiée par rapport au programme initial. Elle fut suivie d'une présentation d'artistes qui avaient été évoqués les jours précédents pendant les échanges directs avec les étudiants.
L'après-midi, explication, recadrage et début de correction individuelle pour les étudiants du groupe pilote et des autres étudiants qui souhaitent présenter leur travail.

J5/8. 28/
Le matin, conférence n°3 : le temps de la peinture (2h30 environ) avec une présentation en trois parties. Cette présentation reprenait le schéma prévisionnel tel qu'il avait été proposé

initialement.
L'après-midi, correction de l'exercice n°2 selon des modalités identiques à la première correction et présentation de l'exercice n°3 pour lequel aucune directive particulière n'est donnée, si ce n'est de rester cohérent avec le choix d'investigation de chacun, et de maintenir bien sûr une relation avec la thématique générale de ce workshop.
L'étudiant est libre et peut éventuellement développer, prolonger les deux exercices précédents, en faire une synthèse ou au contraire aboutir à une proposition nouvelle.

J6/ 8.29/
Repos dominical

J7/ 8.30/
Le matin, dernière conférence présentant des œuvres d'artistes cités pendant les corrections ou répondant à certaines interrogations des étudiants et ensuite quelques exemples choisis de travaux d'étudiants européens.
L'après-midi, les étudiants travaillent pour le rendu du troisième exercice et nous effectuons encore une dernière correction avec le groupe pilote ou ceux qui le souhaitent. Le rdv suivant est le rendu final du lendemain à 14h.

J8/ 8.31/
Le matin fut consacré à une réunion avec l'ensemble des enseignants et ce fut l'occasion de représenter des exemples très concrets de travaux des étudiants européens significatifs quant aux aspects méthodologiques mis en place pour L'ENSA Paris-Malaquais et l'ENSA de Normandie.

J9/ 9.1/
Le matin, nous avons rencontré les étudiants qui le souhaitaient mais ce fut surtout pour des questions de présentation.
L'après-midi, rendu final à 14h. Il a été demandé aux étudiants ou groupes d'étudiants de trouver une présentation aussi rigoureuse que possible et mettant en valeur le dernier travail afin qu'on puisse appréhender facilement leur démarche en rapport avec leur(s) réalisation(s). Certains présentaient l'ensemble de leur production pendant le déroulé du workshop.
Pour cette dernière correction, il est choisi un peu plus d'une quarantaine de travaux et cette fois, c'est vraiment l'ensemble des enseignants qui participe à ce choix de manière à sélectionner ceux qui nous apparaissent vraiment les meilleurs et les plus pertinents.

2.3 Les contenus des cours magistraux

Dans ce chapitre, nous commençons par introduire les conférences prévues telles qu'elles furent construites à partir d'un plan en deux parties. Ensuite, nous indiquons dans la logique du déroulé précédent la manière dont les contenus ont été dispensés en fonctions des cours théoriques mis en place pour le workshop de l'Université du Sud-est de Nanjing.

2.3.1 Cours 1 : Un point dans l'espace

Ce premier cours était déterminant. Il insufflait l'esprit et posait la base théorique essentielle de l'ensemble du Workshop en abordant la représentation bidimensionnelle occidentale. Dans un premier temps, il abordait bien sûr prioritairement l'invention de la perspective centrale au moment de la Renaissance Italienne, mais devait aussi préparer les étudiants à mieux comprendre certaines évolutions et à regarder d'un œil plus averti et plus attentif les modes de représentation d'aujourd'hui.

Plan conférence
1ère partie
1/ Premiers repères
2/ De Giotto à Matisse
3/ La Première Renaissance
4/ La Seconde Renaissance
2ème partie
5/ *Le dessin à l'œuvre* (titre d'une exposition de dessins de Giacometti au Centre Pompidou en 2001)
6/ *Comme le rêve le dessin* (titre d'une double exposition au Louvre et au Centre Pompidou, 2005)
7/ De la profondeur historique au plan contemporain
Ce cours était aussi le premier contact avec les étudiants chinois de l'école de Nanjing et il nous fallait être très vigilent quand à la réception des idées exprimées et leur bonne compréhension.

Aussi, la première image de cette conférence est un petit dessin exécuté avec un feutre sur une serviette en papier dans un bar de Shanghai en septembre 2009 par Jean Li (Xiang Li), architecte et ancien étudiant de l'ENSAPM. Il voulait ainsi résumer une différence fondamentale entre la représentation chinoise par l'utilisation d'un signe et la représentation européenne et occidentale la par l'utilisation d'un petit dessin en perspective. On peut d'ailleurs noter qu'il avait spontanément choisi un chapeau haut de forme occidental pour illustrer sa représentation chinoise et un chapeau chinois pour illustrer sa représentation occidentale. Ce faisant, il notait avec humour la complexité des échanges culturels et nous plaçait au cœur de notre démarche pour ce workshop.
A plusieurs moments pendant cette présentation, nous avons abordé quelques exemples de peinture chinoise et celui un peu particulier de Giuseppe Castiglione, peintre jésuite italien, actif en Chine au XVIIIème siècle.

Le choix du titre fait bien sûr directement allusion au point de fuite de la perspective centrale mais aussi très directement au point de contact qui s'opère entre l'instrument (une pointe de crayon par exemple) et le support (la traditionnelle feuille de papier) dès que commence le processus de la mise en œuvre du dessin. Cela nous permet d'aborder simultanément ce que la langue anglaise nomme *the drawing and the design*. Bien-entendu, on peut aussi le penser avec un logiciel informatique et le trait filaire qui se détermine à partir d'un point dont on a définit les coordonnées.

2.3.2. Cours 2 : Une exposition autour d'une œuvre de Giulio Paolini, De la profondeur historique au plan contemporain (dernière partie du cours n°1)

Ce deuxième cours, qui n'était pas initialement prévu sous cette forme, fut certainement l'élément déclencheur pour la réussite des passages entre apports théoriques et apports pratiques. La veille, notre observation pendant l'après-midi, nos conversations avec toute l'équipe enseignante pendant la soirée nous avait conduits à l'idée d'assurer un cours théorique chaque jour (il était prévu un sur deux) de manière à réduire le temps des interventions. Ce deuxième cours fut bref (40 minutes) et plongeait directement les étudiants dans l'atmosphère d'une exposition temporaire au Musée d'Art Contemporain d'Amsterdam en 2002.
La première œuvre présentée s'intitulait, "*Opera Autentica*", de Giulio Paolini. Elle avait été réalisée en 2001 et son achat par le musée était à l'origine de cette exposition. En tout, une quinzaine d'artistes nous permettait d'aborder plus particulièrement les années 60 et 70 et d'observer par l'intermédiaire de leurs œuvres, généralement blanches et minimalistes, la sortie progressive du cadre dont nous avions étudié la veille certaines définitions et fonctions. Cette forme de cours avait permis aussi les premières questions que nous allions reformuler et développer dans la foulée pendant les TD de l'après-midi.
A la suite de cette deuxième intervention nous décidâmes avec l'équipe pédagogique de maintenir pendant toute la semaine une alternance entre des cours magistraux assez conséquents et grosso modo, tels qu'il avaient été prévus, et des cours plus légers, relativement improvisés et s'appuyant sur les questions des étudiants, d'une part pour tenter d'y répondre, et d'autre part pour développer des problématiques qui justement apparaissaient dans leurs propres travaux.

2.3.3 Cours 3 : Rien que l'idée ! (La transformation à l'œuvre)

Le principe de ce troisième cours était d'accompagner une réflexion sur la notion d'œuvre d'art et les changements radicaux intervenus au XXe siècle mais en tenant compte pleinement des évènements qui anticipent ces changements dès le XIXe siècle et notamment l'apparition des premiers musées.

Plan conférence : Rien que l'idée ! (La transformation à l'œuvre)

1ère partie : Les chefs d'œuvre et le musée
1/ La ronde des chefs d'œuvre
2/ Ordre dispersé, héritages et successions

2ème partie : Art et techniques, frontières ouvertes
3/ Quêtes d'absolu, rapts et disparitions
4/ Copier-coller ou Culture et communication

Cet accompagnement (ce survol très rapide) de la pensée artistique occidentale sur deux siècles devait aider les étudiants à se positionner ou tout au moins à prendre part au débat qui interroge aujourd'hui, d'un point de vue théorique, des notions d'historicité, de modernité, de contemporanéité et d'un point de vue pratique, des approches techniques traditionnelles, des approches technologiques innovantes et des hybridations de plus en plus fréquentes qui associent les deux précédentes. Il s'agit de proposer quelques éléments pour un débat et développer une pensée critique sur l'actualité artistique d'aujourd'hui et l'évolution de son implication sociale et politique face à l'actuelle mondialisation.
Cette approche a été construite en 20 cours d'une heure chacun pour L'ENSA de Normandie. Elle fut reprise dernièrement pour la Faculté d'Architecture de Cagliari en Italie sous une forme comparable.
Par contre, il fallait la repenser complètement sous une forme abrégée et très simplifiée pour ce workshop à Nanjing. Il fallait bien sûr aussi s'articuler avec les interventions précédentes.
La première partie commence avec la constitution du Musée du Louvre et plus particulièrement par la réouverture de la galerie des antiquités en 1803. Issue de la période révolutionnaire (1793), la vocation laïque et encyclopédique de ce musée reste aujourd'hui très vivante.
La deuxième partie du cours était plutôt construite autour des relations interactives qui vont se développer entre l'art, la science et les grandes inventions d'un siècle qui produira notamment les premières expositions universelles.
Le très grand nombre d'informations nouvelles a un peu submergé les étudiants. Toutefois, à partir de ces informations reçues (tous les noms importants étaient donnés en début de séance), à partir des photographies prises pendant le déroulé du cours, à partir ensuite de certaines recherches individuelles sur internet en fonction de leurs motivations personnelles, un certains nombre d'entre eux ont su peu à peu dégager des problématiques pertinentes.

2.3.4 Cours 4 : Rien que l'idée ou la

transformation à l'œuvre (suite du cours n°3)

Ce cours initialement rattaché à la présentation de la veille fut repensé de manière autonome et tient compte des premiers résultats concrets mis en œuvre par les étudiants. Donc après avoir exploré le XIXème nous sommes entré de plein pied dans le XXème siècle.

Nous avons commencé par une grande ballade dans le musée d'art contemporain de Washington à partir de photographies que j'avais réalisées en 2002 pour mes cours à L'ENSA Paris-Malaquais. J'avais pris soin de photographier une série d'œuvres, principalement des tableaux des années 60 à aujourd'hui, de face bien sûr mais aussi de côté pour montrer toute cette perception modifiée de la peinture que peut entraîner un point de vue dynamique (en mouvement). Ainsi nous prenions également la suite du cours précédent, lié à l'exposition d'Amsterdam (œuvre de Giulio Paolini) qui, elle, avait été photographiée de manière classique.

Ensuite, nous nous sommes concentrés sur une période qui commence par la mort de Cézanne en 1906 et sa rétrospective en 1907. Nous avons étudié l'évolution des rapports entre la figure, la forme et le fond notamment avec Matisse et Picasso, puis la disparition progressive de la figure et la naissance de l'abstraction telle qu'on peut l'observer chez Mondrian et Malevitch. Nous sommes restés plus particulièrement sur les années allant de 1906 à 1930, déterminantes notamment à Paris qui à cette époque est un foyer culturel international de premier plan. Enfin, nous sommes allés saluer Marcel Duchamp...

Nous nous sommes aussi intéressés aux nombreux moyens actuels de duplication, mécaniques (Walter Benjamin) mais aussi bien évidemment technologiques.

Nous avons ponctué ce parcours par un petit panorama d'images présentées sous forme de « vignettes » sur un seul document et qui permettait l'observation des rapprochements possibles entre toute une série d'œuvres couvrant une période de plus de 150 ans.

Enfin pour terminer, c'est lors de cette matinée que nous avons présenté pour la première fois aux étudiants chinois quelques exemples de travaux d'étudiants français qui avaient eu l'occasion de répondre à des exercices différents mais comparables.

2.3.5 Cours 5 : Le temps de la peinture

Cette troisième conférence s'inscrit elle aussi dans la logique des deux précédentes, mais cette fois la cohérence passe par mon travail personnel et plus particulièrement par ma peinture. Des rapprochements seront faits avec des artistes, des œuvres étudiées précédemment mais aussi avec mes propositions pédagogiques.

Plan conférence : Le temps de la peinture
1ère partie
1 / La profondeur d'une surface
2 / Une appartenance européenne
3 / Un apprentissage moderne
4 / Un apprentissage classique
2ème partie
5 / Carré divisé ou double carré
6 / Polyptyques
7 / La revendication du dessin
8 / Le peintre et l'architecte

Ce cours fut présenté tel que nous l'avions programmé puisqu'il était de nature un peu différente et conçu à partir d'une conférence sur mon travail que j'avais eu l'occasion de présenter en 2009, respectivement à Dalian et Shanghai. Tout en présentant un déroulé thématique de ma production artistique, cette conférence, par le jeu des références, permettait aux étudiants de retrouver des œuvres importantes, initialement présentées dans leur contexte historique et qui étaient cette fois, de manière plus subjective, présentées par association, par confrontation, avec les hommages ou autres développements qu'elles avaient entraînés dans ma pratique personnelle.

Par ailleurs, ce dernier cours, à la veille du week-end, informait aussi les étudiants sur mon propre positionnement artistique et rendaient peut-être plus clair et lisible l'ensemble des cours, à la fois sur leur "objectivité historique", mais aussi le parti pris choisi pour nourrir la pédagogie de ce workshop.

2.4 Les réactions des étudiants et le dialogue

Le cours théorique d'introduction du premier matin fut suivi avec une grande attention mais il y eu ensuite un moment de flottement pour beaucoup d'entre eux. Cela a duré toute l'après-midi de ce premier jour, temps pendant lequel les étudiants du groupe pilote, ou des autres groupes ensuite, venaient demander régulièrement des explications supplémentaires et même en quelque sorte les réponses qu'ils étaient censés chercher et trouver eux-mêmes à partir de l'énoncé du premier exercice.

Pourtant, on pouvait sentir que d'une manière générale, beaucoup d'étudiants avait fait preuve d'une appropriation assez réactive et positive aux cours théoriques et aux images avec d'ailleurs dans la foulée des recherches appropriées.

D'où l'idée, au TD du deuxième jour, de partir de leurs réceptions sensibles aux deux premiers cours théoriques par ce que nous avons appelé "l'expérience des trois mots" et qui consistait très banalement et sans trop réfléchir à écrire les trois mots qui leur semblaient d'une manière spontanée le mieux correspondre

aux impressions les plus fortes qu'ils gardaient des images, des informations ou des idées exposées précédemment. Parfois ce fut un seul mot mais on faisait avec.
J'ai pu alors trouver assez facilement, un peu à ma surprise, un contact assez vrai avec les étudiants, quelquefois directement en anglais, mais la plus part du temps en passant par le chinois et le français grâce à Wang Ying qui trouvait très vite les mots justes dans les deux langues.
Un nombre significatif d'entre eux se sont alors appropriés "intimement" les exercices et ont joué le jeu avec une certaine liberté, très clairement encouragée par Monsieur Zeng.
Puis, l'exemple de premiers résultats significatifs en a encouragé beaucoup d'autres et à partir du groupe pilote nous avons réellement pu travailler avec l'ensemble de la promotion.

Ensuite, tout va très vite (parfois presque trop) avec une rapidité d'exécution et des moyens vraiment très diversifiés. Il y a eu de nombreux objets dont certains furent très bien aboutis dans leur réalisation technique. Il y a eu quelques installations, mais là effectivement, le manque de temps ne permettait pas en général un réel aboutissement. Il y a eu de l'humour, de la pertinence dans certaines réponses et beaucoup d'idées. Plusieurs fois, il y a eu l'usage du feu, ce que je n'ai jamais vu avec les étudiants européens.

2.5 Le dialogue avec les enseignants chinois et réajustements éventuels

Il a été permanent, régulier et progressif. Aujourd'hui encore, nous continuons dans ce sens.
Or, nous l'avons vu précédemment, un des attendus importants pour l'Université du Sud-est, est de mettre en place un apport méthodologique et sa bonne coordination avec les enseignements existants. C'est donc sur une cohérence liée à cet objectif que nous nous rassemblons :
-Cohérence entre les propositions d'informations apportées par un enseignant européen (qui ne parle pas le chinois) et leur assimilation potentielle et réelle par les étudiants chinois.
-Cohérence avec les propositions qui vont être transmises par mes collègues chinois à partir de quelques exemples plus directs avec les étudiants.
-Cohérence entre un workshop de 10 jours et un enseignement annuel avec une observation sur le moyen ou le long terme de la part de l'Université du Sud-est.

3 Analyse des travaux

Tout d'abord et de manière générale, les étudiants ont investi ces exercices avec beaucoup d'attention et de bonne volonté.
Ensuite, progressivement pour un bon nombre d'entre eux, ils ont poursuivi avec une certaine liberté. On précisera que le dialogue fut constant avec les étudiants pendant toute la durée du processus, mais qu'il ne fut pas demandé, dans le cadre de ce workshop, d'accompagner les réponses par des textes, afin de ne pas alourdir la démarche en cours par tout le travail de traduction, que cela aurait induit forcément. Ce sont donc bien directement les propositions plastiques qui ont prévalu pour notre choix.
Toutefois, en accord avec les enseignants de l'Université et avec l'aide de Wang Ying, tous ensembles, nous avons sélectionnés une quarantaine de travaux des étudiants chinois, selon des critères assez comparables à ceux des étudiants français montrés au préalable (chapitre 2.4.1 & 2.4.2). En effet, même si les intitulés des exercices étaient un peu différents (chapitre 3.2 J1 & J3), la qualité d'aboutissement d'une mise en oeuvre, la poésie, l'esprit de finesse, la simple évidence quelquefois, furent les éléments déclencheurs de nos options. S'ajoute à cela un autre critère, très important dans le processus du travail quand les étudiants sont réellement et authentiquement investis, celui qui fait apparaître une continuité et peu à peu une cohérence dans l'ensemble de la démarche.
Ainsi, progressivement, d'un exercice à l'autre, se font jour des préoccupations propres à chaque étudiant et c'est là aussi le caractère sensible de cette approche que nous essayons de sauvegarder et de vous présenter. Il ne s'agit pas de vouloir à tout prix être original, mais simplement d'être sincère et vrai dans ce que l'on fait.
Ce fut vécu, parfois avec enthousiasme, par celles et ceux dont vous découvrirez les travaux, et nous espérons que pour l'ensemble de la promotion, l'expérience comptera.

3 作品分析

首先，总的说来，学生们对这些练习给予了足够的重视，并投入了大量精力。其次，对于其中很大一部分学生来说，他们在研究过程中慢慢地融入了一定的自主性。需要指出的是，每个过程从头到尾，教师都与学生保持了良好的沟通，但整个教学活动并未在框架内提出作品需附有文字说明的要求，为的是不加大翻译工作量，从而拖累整个进程。因此，真正直接影响我们选择的是造型设计。

然而，在东南大学教师们的配合下，在王盈及所有人的帮助下，我们按照之前向大家展示的与法国学生类似的标准，选择了40个中国学生的作品。事实上，尽管练习的标题有所不同，但作品制作的质量、诗意、幽默，有时还有简洁明了，是我们做出选择的关键因素。此外，加上另外一个非常重要的标准，在工作过程中，如果学生真正全身心投入，还应在作品制作过程中突出连续性以及整个步骤之间的连贯性。于是，从一个练习到另一个练习，渐渐地，每个学生开始有了各自新的关注点——这也是我们想保留住，并想呈现的，有关于这种教学方法的独特的感性特点。这并不是说不惜一切代价来呈现新颖，而仅仅是对于我们所做的作一次真诚的、真实的记录。这就是真实的经历，有时，通过一些学生的作品，我们体验到创作的激情。有时，对于有些学生作业的结果，所表达的准确度还不到位。但不管怎样，对于所有的学生，最重要的是这份经历。

3 Analysis of work

First and generally, students have completed the exercises with great attention and goodwill. Then, gradually quite a few of them continued to work freely. We need to specify that the dialogue with the students was constant throughout the process. However, they were not asked in the context of this workshop, to give explanation in the form of articles, in order to avoid the translation work which could have complicated the process. Thus, the plastic proposals immediately became our preferred choice.

However, in agreement with the teachers of the University and with the help of Wang Ying, we have selected some forty works of Chinese students, according to criteria quite similar to those used for the French students shown previously (Chapter 2.4 .1 & 2.4.2). Even if the titles of the exercises were a bit different (Chapter 3.2 J1 & J3), the qualities of implementation, poetry, humour, sometimes the simple straightforwardness, were the triggers for our choices. Added to this is another criterion — very important in the work process when students are really and truly involved — the gradual appearance of a continuity and consistency throughout the process. Thus, gradually, from one exercise to another, emerged concerns specific to each student — this is also the sensitive nature of this approach that we try to preserve and introduce to you. We do not want to be original at all costs, but simply to be sincere and true in what we do. You are about to discover the work of those students who have - sometimes enthusiastically - experienced this. We hope that this venture will make a difference for the entire class.

案例 1

翟炼

作业 1　没有什么比一个表面更有深度

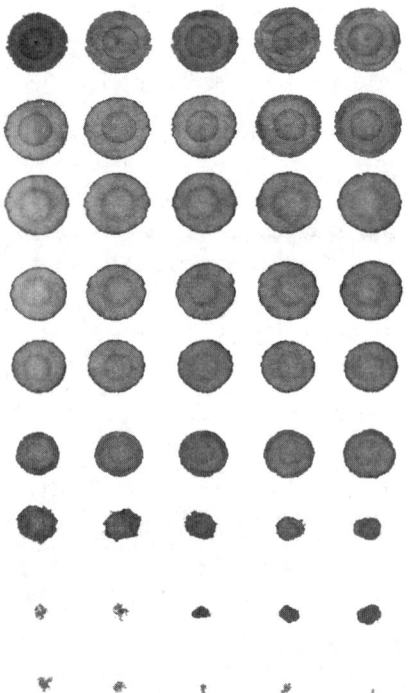

表面：事物的非本质属性，即表象。
深度：事物的本质属性，即内在。
两个哲学层面相悖的名词，在自然界中却总是完美的统一在一起。
表面总是通过一定途径达到深度的本质，或距离，或光影，或空间……
表面自有深度，因为这是个现实的世界。

The surface : Extrinsic attributes of things, the idea.
The depth : Essential nature of things, the immanence.
Two words which mean opposite in philosophy, join perfectly in nature.
The surface always reaches the essence of depth through some ways, such as the distance, the shadow, or the space.
The surface has its own depth, for this is a reality world.

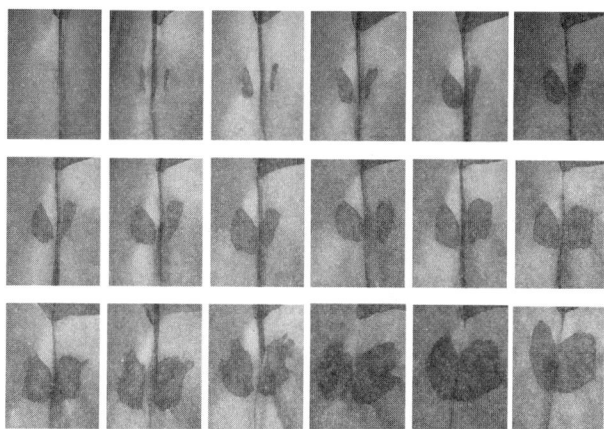

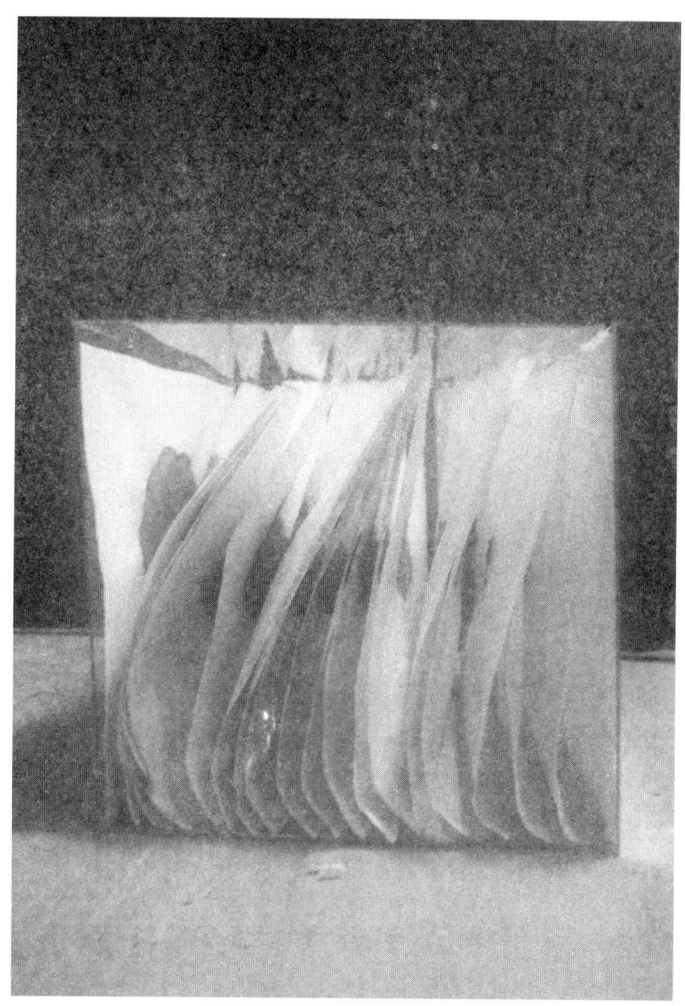

一个墨点　　　　　　　　A blot
从它降生的那一刻　　　　From the moment of its birth
就在努力渗透着，扩散着　Struggles to infiltrate, to spread
它的绚烂，美如夏花　　　His gorgeous, beauty as the summer
……

终于，在一张纸的面前　　At last, in front of a piece of paper
它停下了脚步　　　　　　He stops
累了，倦了　　　　　　　Tired and weary
那就停下来吧　　　　　　So just stop
这个历程　　　　　　　　This process
就是生命的深度　　　　　Is the depth of life
……

作业2　建筑是所有运动的框景

一个圆洞
照壁？空间？
或者，更像是一轮太阳

A round hole
The screen wall? The space?
Or, it is more like the sun.

走近，走近
侧面窥去
原来
是弯新月

Approached, approached
Peek by its side
So that
It is a new moon

建筑
是一种诠释
诠释诗意的自然
以让充满劳绩的人们
栖息心灵的家园
……
框景
是一双眼睛
缤纷斑斓
只有契合魂灵的美丽
方有栖息的恬然
……
是墨点
打开它
就是打开了一扇门，一面窗
静静守候
这宁静里
有你想要的悠然

Architecture
Is a kind of interpretation
Explain the poetic nature
To let the industrious people
Habitat at the home of heart
…
The frame
Just like eyes
Colorful and gorgeous
Only the beauty fits the heart
Has the leisure of habitation
…
A blot still
Open it
You open a door , a window
Wait there quietly
In this quietness
You can find the leisure you want

作业 3　生命

是晕开的一瞬	The moment of its spread
是渗下的一抹	The point of its penetration
是凝成的绚烂	The joint of its gorgeous
是余下的空白	The blank of its left
……	…
我来过	I came before
看到那美丽的绽放了吗	Have you seen the beauty of my bloom?
我爱过	I loved once
听到那静静的倾诉了吗	Have you heard the silence of my whisper?
……	…
终于	At last
我还是消失了	I disappeared
……	…
白纸	The blank paper
无穷无尽	Endless and spotless
最美好的	The most beautiful things
是那些印记	Are those memories
因为	Because
那是我的生命	That is my life

案例 2

姚曜

"线的图像"体现在一张用铁丝弯成的人脸上,这里图像是重点,线是组成元素。"图像的线"体现在后面飘起的头发上,叶片暗示了头发同时也是吹拂叶子的阵阵清风,这里,线是表现的重点。以图像为载体,线索的曲直硬软,力量与韵律,为作品的形象进行二次塑造,甚至取代图像成为形式的主导,表达作品深层次的情感与韵味,同时传递出更多的隐含图像。这幅作品中,我们似乎看到:"轻拂湖面的风姑娘"、"绿野中奔跑的少女"、"飘逝而去的青春"……这里图像与线的关系不再只是整体与元素,他们彼此互相塑造交相映衬,给观者更多的解读的空间。

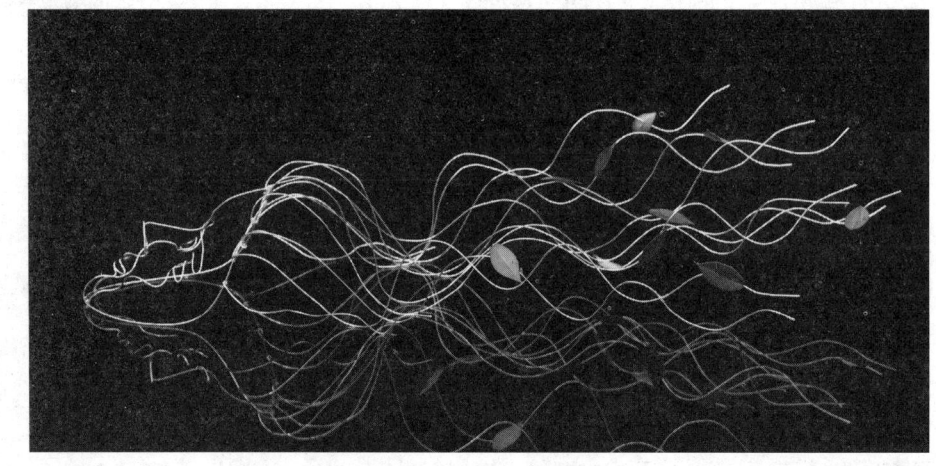

案例 3

孙志峰

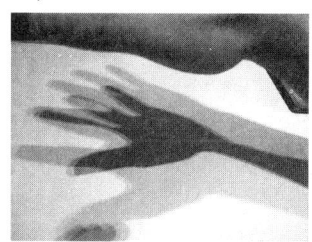

主题：大与小，内与外的变化

题目1：没有什么比平面更有深度

作品：

（1）以桌面为画布平面，利用不同光源对手投下的阴影产生叠影和不同灰度，光影暗示了画面之外光源的方位，形成空间的深度，以小见大。

（2）利用中国刻章的方式，用卷纸在画布表面密布敲章，以小的图案单元形成大的平面，其中暗藏这卷卷纸，卷纸可以下按形成凹尖或向上凸成锥体，体现深度。

（3）利用相对大小，远看是很细微的东西近看未必。纸这个材料，远看是个平面，近看表面的细孔却相对放大，可以透过看到背后的景物，体现深度。

思考：

（1）（以小见大）表面和深度，想寻求画面之外的深度，暗示画外的空间，于是寻找画与画外的联系，想到了光线，利用画面暗示光源的位置，暗示空间。

（2）（小聚成大，大中见小）联想小时候玩的胶带圈可以摁下可以挤出，这是一个可以体现空间和深度的玩意，于是考虑用此为材料创造一个表面，这个表面由圈的图案组成，考虑用刻章的方式。

（3）（大小相对）联想到事物宏观和微观的区别，很多在宏观上都是简单的表面，而在微观层面有自身的结构和深度，像皮肤等。但是皮肤难以表达，于是我想用其他的物品替代，比如布或纸，同样有两层且更容易表达。

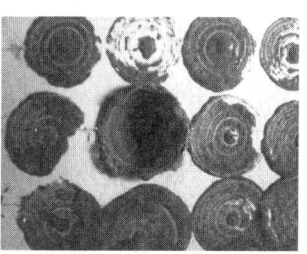

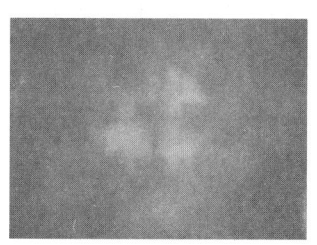

题目2：图像的线条，线条的图像。

作品：
　　画面中长出的三维的树，树从画中来，又在某些地方进入画面，完成二维平面与三维空间的融合和转换，同时体现二维和三维上线条的表达。

思考：（亦内亦外）
　　线条与图像，由线条组成的图像，希望图像和线条同时保持自我，选择有张力的线条是必要的，想到枯树，感觉挺好，特别是那种遒劲的枝干，线条本身就有种力量感，不会轻易失去线条本身；画布和树穿插，形成交融，二维和三维交融，亦内亦外，最后抽象出树枝的线条，形成水平和竖直的构成。

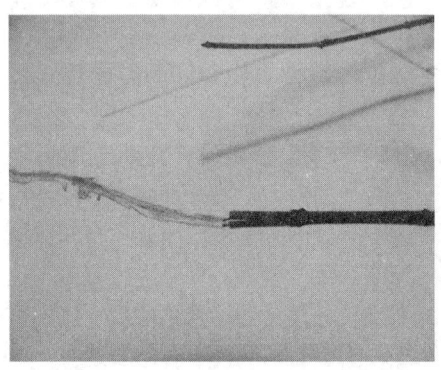

题目3: 自拟

作品:

(1)(利用小装置反映时间、空间、人的关系，表达时间空间和人的交融，在时间中创造人处的空间，在空间中延续人处的时间。

(2) 延续第二个设计，树作为时间在空间层次上的表达，暗示时间的流逝过程，由树枝抽象出的线条看作时间树在画布上的投射，在画布上形成的时间刻度，同时纵横的线条让人联想起城市的街道，而街道与时间紧密相连，有它的发展过程，从而构成完整的表达。

思考:

(1)（以小见大）空间绘画的启发给了增加维度的可能性，使可以表达抽象的生存状态。

(2) 第二个设计中的抽象线条可以有多种解释，结合(1)中的时间概念，树可以表达时间的概念，于是有了这个设计。

案例 4

查琳

For our short term of the third year of the architecture course our class attended a class given by a French arts teacher named...

From the PowerPoint presentations we could learn from artworks made by historic artists and also modern ones either professional or students.

We could see artworks both concrete and abstract from very simple to complicated.

Some of the artworks are even seemingly controversial.

After the lecture, we went to take one-on-one classes in groups with our French and Chinese teachers.

Our foreign teacher told us that we were going to take part in a project where the assignment would come from within.
I noticed that in the beginning it caused a bit of confusions, because we were lost in the lack of boundaries and we were not told exactly "what" to do.
We were totally free in the production of our idea, of which at the time we had no clues what it was.
No instructions for dimension, material use, scale, abstract, concrete or what so ever.
It was "our individual problem" as he said that what we were going to do and how we were going to do it.
We basically had to reach inside of our selves and look for inspiration.
To guide us through finding the centre for our piece, the teacher asked us to think of 3 words that we would like to express.
These 3 words would be a starting point that could change into ultimately one word or even six in my case.

For me coming up with an idea was quite easy, because I have always been amazed by the 3 things that are the most mysterious to mankind — being time, space and energy-..
Then we had to write a short introduction about it that I ended up with the Big Bang Theory where my idea Ultimate Creation, Destruction and Recreation into Infinity came to mind.
With that a scratch..
To my surprise the teacher had little comment and supported my idea before I explained.
The next step was to work on a concept of the first piece.
I went back to the workshop and tried to refine the rough scratch, which I thought I did in my sketchbook.
The more I worked on it, the more the idea expanded.
The next day I returned to my teacher to show him what I had come up with.
From his first reaction I had the impression that he did not like it at all.
He asked me for the original sketch, but I did not carry it with me at the time.
He noted that too many artists find themselves drawing away from the original raw emotion given at the moment of the inspiration in attempt to create their work.
I really did find myself in that position and had to rewind so I could come up with something better.
He advanced me to make my original drawing and better to use coal sticks instead of colors
Second time around at the studio I got a big sheet of paper and started working on it.

The goal was to try to express the picture in my mind of the occupied universe that we were familiar with in alignment with other universes that were even more mysterious.
I had the idea that at a certain point in time a lot of energy was released out of the centre of the "mulitiverse" and created everything.
Ultimately, occupied space would reach it's maximum potential and eventually went back into the centre and destroyed itself.
In the recreation the same thing happened all over again, but just a bit better than the previous time in the sense of eternal evolution.

With this in mind, I colored the sheet black from the centre on out in a rotating fashion, resembling infinity in space.
Then I started drawing circles which stood for the occupied space in the multiverse, each resembling another kind intertwined with the other universes.
I, myself, was at ease with the result, which thankfully was also shared by others, including my teachers.

In the next phase of this project we had to come up with something completely new or a production which would either complete or extend the first.
Just like the teacher said, it was in my experience much harder to come up with something better when the previous one was already satisfactory.

For this artwork we were given relatively more time to come up with a presentation.
In my case I had no idea what to make it better in the first place. After sleeping on it I got an idea the next morning of making my 2 dementional drawing into a 3 dimensional lamp.
In the centre of the lamp there would be the actual bulb standing for the centre of the multiverse with

spirals increasing in diameter intertwined with one another were going to mage an engaged space

If the licht were to be turned on in a dark space, then we could see engaged space where everyting was created.

When the lights would be turned off, it would all metaphorically disappear into the centre of where it all started.

By turning the light on and off it would somehow express eternal repetition

I came up with a maquette the next day and showed it to the teachers. I didn't have to use many words to express what I was trying to pull off. He was fond of the model and I just had to find a way to make it on a bigger scale.

I went home and started working on the wires because it had to be on a bigger scale, I wanted to wind the wires up so that it could become stronger.

I ended up spending a lot of time twisting the wires and I chose the green color because it represent life to me. Since the objects in space were in constant motion, it was to some extent "a living everything".

However, the time I had spend in attempt to execute my final work compared to the time there was left to actually experiment with a new material was not balanced in my favor.

I was actually disappointed from the time that I realized that the material was not suitable for what was the intension of the expression.

This experience was also shared by my teachers.

All in all the most important lesson that I've learned from this project is that as a creator of designs, artists have to stay true to the core of their inspiration.

It is how ever necessary to find a balance between that Upgrad the idea and ultimately find the right material that will help embody the desighner's vision.

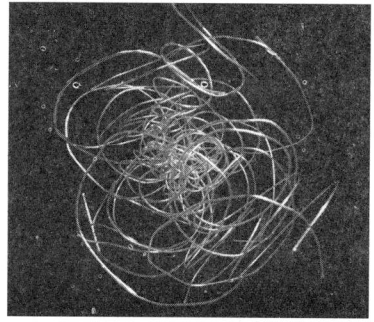

案例 5

刘晓帆

盆栽的国画
Potted Painting

案例 6

张钊 彭文哲 钱凯

"没有什么比一个表面更有深度。"
"所有的建筑都是对运动的一个框景。"
即使你不能看见,也并不意味着它部存在。
由一堆工作室废弃的垃圾所搭成的貌似非主流的作品,经过灯光的照射,在白幕上投射出类似一座城市的剪影,就像是小时候玩的影子游戏一样。

Nothing can be deeper than a surface.
All the architecture is a scape frame to the movement.

 Even you can't see something, does not mean it isn't there!
 Put up by a pile of discarded rubbish from studio which looks like non-mainstream works. After light irradiation, project the silhouette of a city on the screen, just like the shadow games played during our childhood.

案例 7

施婧

作业 1　没有什么比表面更有深度

用三个词组来阐释这句话，即简单与复杂、平面与立体、叠加与转化。

在初始阶段，我认为点是最简单的却可以组成复杂的画，这体现了简单与复杂，而复杂的画中的点若被按排折起来，透过透明的纸张，看到的将是一排点，但此时纸不再只是平面而具有了厚度，在平面与立体的转化中图像也随之变幻，并且这两个动作中均体现了第三个词组，即叠加与转化。

但在后来的揣摩中，我发现这个作品中不可忽视的是我手上的动作，而在一个静止的作品中我的意图也许很难被理解，且如果能在似开未开时从各个角度发现不同的景象应该更妙。

于是，我选用了半透明的硫酸纸，并重新选取由点组成的图像，即两张看起来类似却截然不同的脸，从一个角度看到的是哭脸，而从另一个角度看到的是笑脸，但两张脸均是哭中带笑、笑中含哭，充分体现了矛盾性与融合性。

作业 2　三线相交

在第二份作业中我决定延续第一次作业的两面性与复杂性。

我用具有反光性的镜子代替了硫酸纸，将镜子割成一片片并将它们之间的夹角呈160度左右排列。在考虑镜子之前的图案时，由于镜子呈锯齿状，如果是复杂的图案在镜子中只会变得更复杂，同时发现镜中的影像有时会重合。我开始思考怎样利用这种特性：我试着将一串点与镜子相交，发现在镜内影像的左右侧各出现了一条点，形成三线相交的景象，但因为点都是完整的，因此线的端头并未相交，在将端头的点改成120度的扇形后，便成了完整的三线交汇。

作业 3　标志

出发点是一种幽默的态度，想做出一些标志，既是作品，也是对最后布展时展厅氛围的一种营造。

主要手法是通过打孔，用空的圆点形成图像从而产生一种朦胧之美。远观是一个整体，近看却有不同感受，延续了两面性与复杂性。

最后作品分为两个，一个从左和从右看各是向左转与向右转，另一个则是禁止车辆通行标志和禁止吸烟标志。

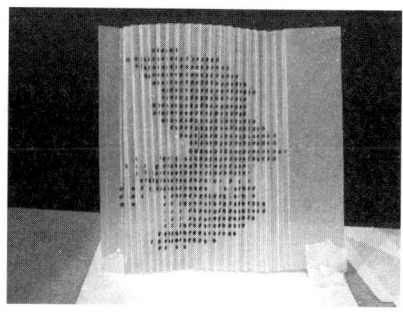

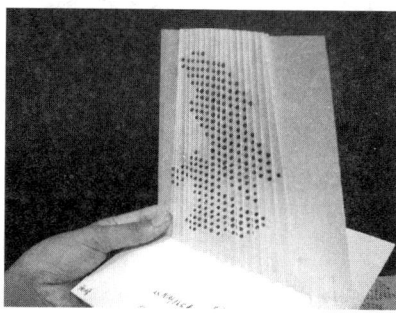

Job one: There is nothing deeper than a surface.

From my point of view, there're three phrases to explain this sentence — dimension, superposition and transformation.

In the first stage, I had a thought that a point was simple but could form a complex picture, which reflects the simple and complex; and if points of a complex picture were folded according to row, through the transparent paper, we would see just a row of points. In this time the paper is no just two-dimension but three-dimensional. With the transformation between the plane and three-dimensional images also would be changed. The two movements both reflected the third phrase — the superposition and transformation.

Then in the second stage, I found an important part in this work both is the actions of my hands, but in a static work, my intention may be difficult to be understood. And if you can not open when you open like from all angles different picture that should be even better.

So, I chose the translucent sulfuric acid paper, and re-select the image formed by points which looks like two very different faces. From one angle is a crying face while from another angle is a smiling face. The two faces both have smile and tears, fully emboding the contradictions and integration .

Job two: Three lines intersect

In the second assignment , I decided to continue operating the two-side feature and complexity.

I used a reflective mirror instead of sulfuric acid paper and cut the mirror into pieces. Each angle between them is around 160 degrees. Then I began to consider the pattern before the mirror. Because of the jagged mirror, if it is a complex pattern, in the mirror it will become more complex. I also found the image in the mirror sometimes overlapped, so I began to think about how to use this feature. I tried intersect a string of points and found

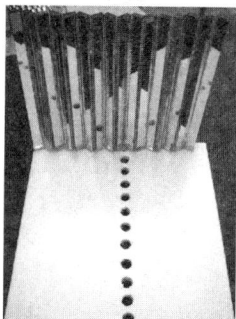

三线相交

a mirror image of the left and right side mirrors the emergence of a point, three lines intersect to form the scene, as the points are complete, the ends of lines do not intersect. In the end, I for point of 120 degrees into the sector after they have become a complete three-lane intersection.

Job three: Sign
The starting point was a humorous attitude, I wanted to make some signs as an assignment and also to create an atmosphere for the final exhibition.
The main way was through punching holes, using the empty dots to produce a hazy but beautiful image. From a distant view, it was a whole while it has a different feeling from a close view. It continued the two-side feature and complexity.
Finally, the work had two parts: in the first work, from the left and right side, you can see a left-turn and right-turn sign, and in the second, there is a traffic sign and a no-smoking sign.

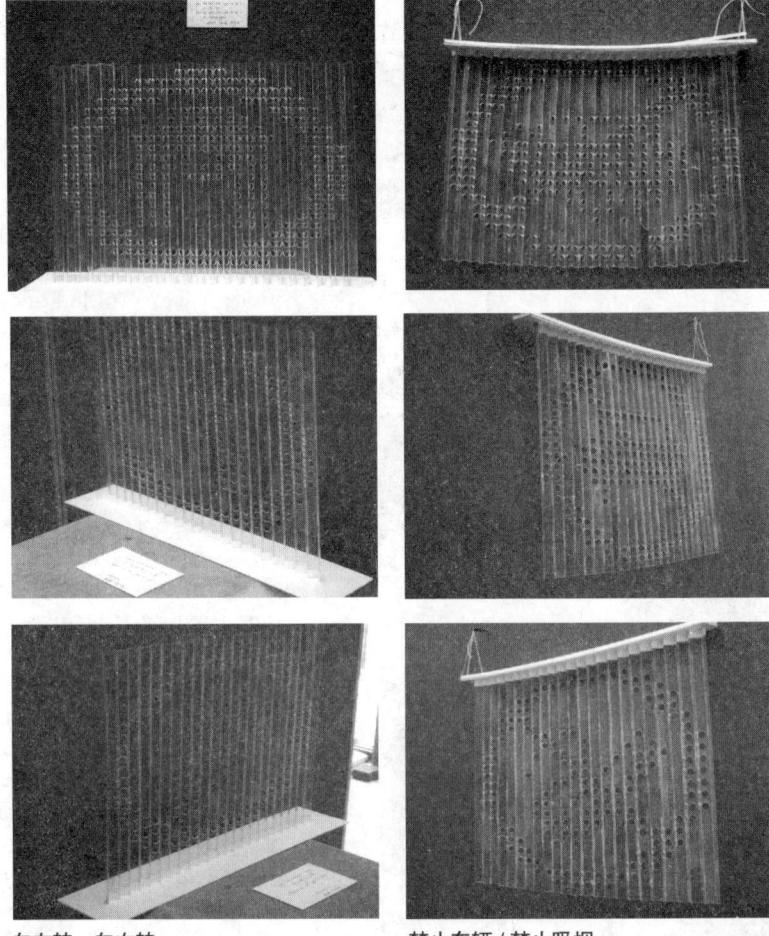

向左转，向右转　　　　　禁止车辆 / 禁止吸烟

案例 8

徐振欢

受马列维奇的作品《白纸的白方块》（图a）的启发：简单两个面叠加，平面图案仿佛产生前后交错存在的空间。

作者将一张白纸揉出机理，用钉字钉在一块木板上(图b)，木板的左上角放置光源，白纸在木板上投射出阴影，因此，富有机理的纸张和右侧的阴影形成错综复杂的明暗关系。

作者用铅笔，在一张白纸上描摹出木板上的纸张和其投下的阴影（图c），这幅画在白纸面上产生悬浮感。同时，作者用真实的银白色图"钉"钉在纸张的上角，以一种幽默的方式表现平面的虚无空间。

The work is inspired by a painting "A White Cube on White Paper"(Figure a) by the famous artist Malevich: two simple planes are put together, thus creating the space between them.

 A piece of paper is rubbed and nailed on a board (Figure b). The light casts the shadow of the wrinkled paper on the board and creates complex relations between light and shade.

Then he draws a picture (Figure c) of the paper and the shade on a piece of white paper, and the paper seems to flow on the surface. Interestingly, he nails two real silver nails on the picture as if a real piece of paper is fixed on the white background, which present space between planes in a humorous way.

图a

图b

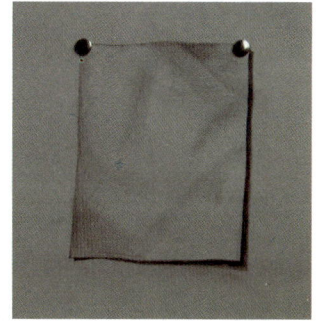

图c

案例 9

余海男

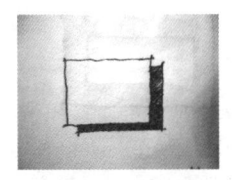
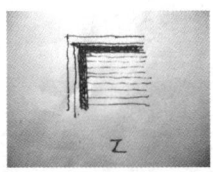
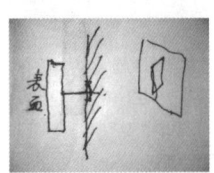
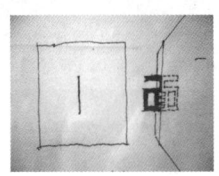

作业 1

无限 二元 交流 — Infinite binary exchange

镜面反映的内容具有无限的深度和广度，借以指代外部世界。进而以为外部世界的像因经过薄薄一层物质表面的转换后为人所接受而更为客观。

半个"面"按文意可理解为描述性的，也可仅被视作一个面状的符号。重要在于它是非客观的，表达了人的智觉因素，借以指代人的内部世界，其含义是无限的。

两个面垂直接触以产生交流。主因客而完整，客为主所侵入：镜面与符号的交流成为客观与主观的交流。

The reflection of a mirror contents unlimites depth and breadth which refers to the outside world. Then the outside world through a thin layer of material is more objective.

Half of the "face" according to its context may be understood as descriptive, but also be regarded as only a "face-like" symbol. The importance is that it is non-objective, expressing the intellectual sense of human beings. Refering to people's inner world, its meaning is unlimited.

The vertical contact of two surfaces makes communication. The subject is completed due to the object. And the object is invaded by the intrusion: The contact of mirror and the cross symbol means the exchange of objective and subjective.

作业 2

"无论孔洞或窗口，墙都相对之构成边框。墙为实，孔为虚。一方面将观察者与观察的对象隔离：为看提供隐蔽；另一方面又要求观察者为安全感付出代价：片段，看不到的剩余。"

"看和被看的经验与看和被看的两个空间。"

——张永和

在前者的基础上，尝试对"窥视关系"的进一步探讨。

偷窥者的困境

体验者进入三联厢的居中的一间内，左右两侧厢壁上都有窥视孔，望进去希望一窥相邻厢房的内容。起初难以辨认看到的内容，当偷窥者意识到观看的只是自己经过一组镜面后呈现的虚像时，体验结束。

看和被看两种经验同时体验，看和被看两个空间重叠。

偷窥者的协约

装置被放置于展室的中央，四面厢壁上都设有窥视孔。好奇的参观者从四面向内看——厢内没有内容物，窥视空房间的偷窥者们"以眼还眼"，静默中履行彼此的协约。

"Whether holes or windows, the walls become border. Walls are real and holes are vain . On one hand, separating the viewers from the observed objects and providing cover for the viewer; On another hand, requiring the viewer to pay the price for their security: fragment ,the invisible rest. "
"Experience to see and be seen and see two spaces to see and be seen."

—Zhang Yonghe

Based on the former, to attempt further study on "peep relations" .
The plight of peeper
 A Experiencer enters the middle car of a triple room with on both side of the walls a peephole. He looks into and wants to have a glimpse of the contents of the adjacent rooms. It is difficult to identify the content at first.Then he awares that he is seeing himself. The experience ended.
Experience to see and be seen at the same time and two spaces overlap.
The treaty of peeper
 The device was placed in the central showroom, surrounded by the walls with peepholes which curious visitors look for the through inside — no contents inside the device, and voyeurs share "an eye for an eye", performing their agreement in silence.

偷窥者的困境

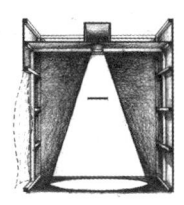 

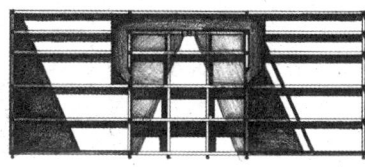

偷窥者的协约

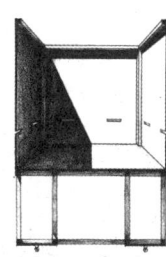 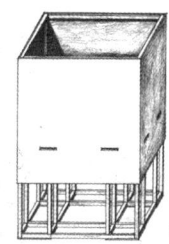

作业 3

运动表明时间向度的存在。时空二者共同构成经验的世界。尝试解构通常性的经验时空，以获得更多可能性。

（1）若在空间中设立一个切面，以表明二维世界的广度，又以不具方向的时间为第三维，即获得这样一种经验：过去，现在和未来全部叠合在薄薄的平面上，二者构成新的三维结构，并获得宏大叙事的全景。

（2）若在空间中设立一个切面，又以具有方向的时间为第三维，即获得三维物体在二维表面上的一系列投形，共同组织起微叙事性质的经验片段。

以" Etienne Jules Marey "的运动影像分析"和杜尚的"下楼梯的女人"作为先例，对第二种时空观进行尝试。

Movement shows the existence of time dimension. Both time and space constitute the experience of the world. Try to deconstruct the experience of space and time to get more possibilities.

(1) If set a section in the space to show the breadth of two-dimensional world, and time without direction as the third dimension can be obtained such an experience: The past, the present and the future all superimposed on a thin plane, forming the new three-dimensional structure,which access to the grand narrative panoramic views.

(2) If set a section in the space, and time with direction as the third dimension, three-dimensional objects in two-dimensional shape on the surface can be obtained, which access to the micro narrative experience world.

Using "Etienne Jules Marey's motion image analysis" and Duchamp's "The Woman Walking down the Stairs" as a precedent, try the second idea of time and space.

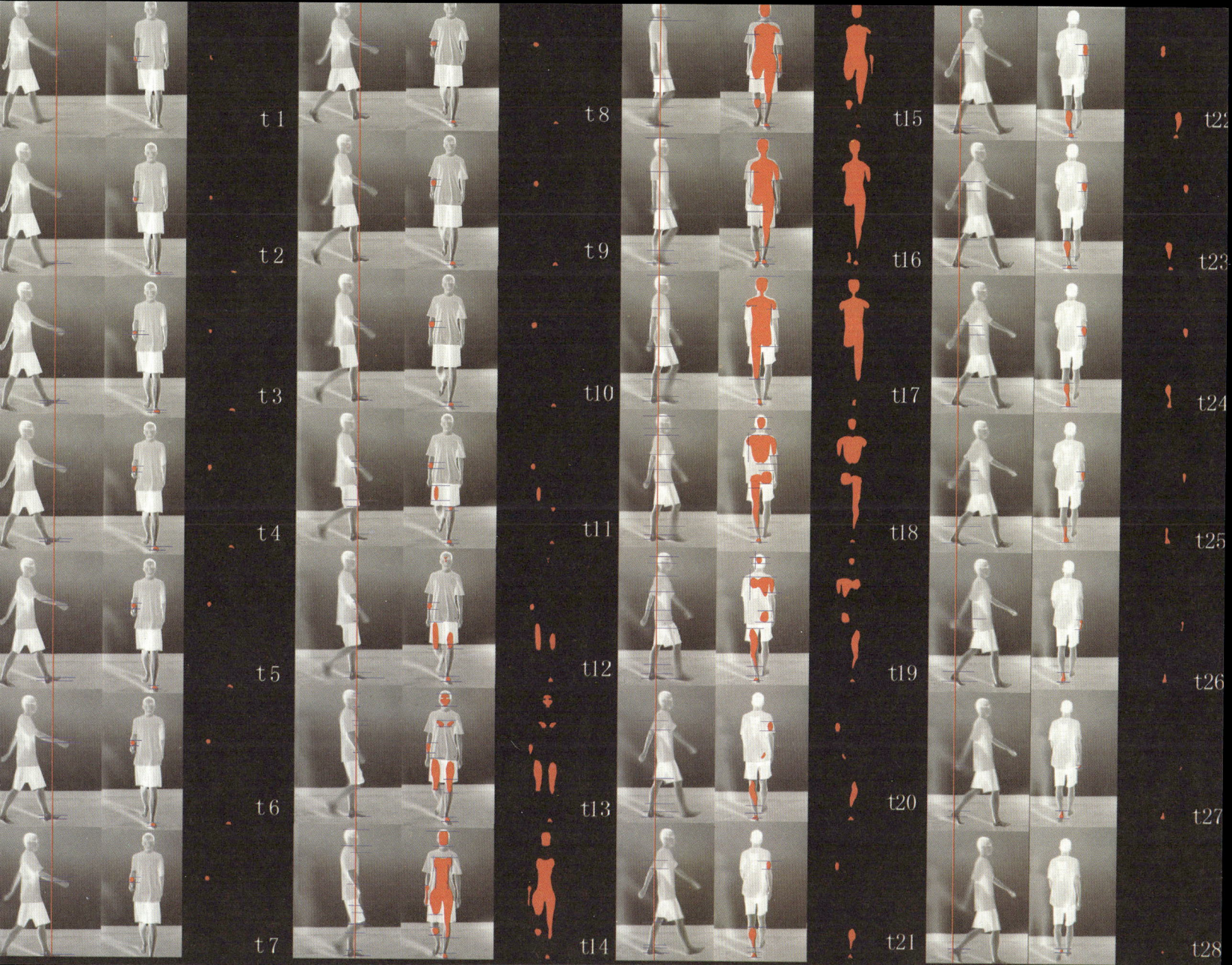

案例 10

孔斌

作业 1 "没有什么比一个表面更有深度"

设想一张画,画上的内容是固定的,而观众在观察它的过程中却感受到了不同的风景。如何实现?利用人的视点会改变的现象,人在绕着画走动的过程中,由于观察作品的角度不同,所看到的风景也不同。作品用白色长条的扭转与旋转来形成有秩序的画面,旋转的处理强化了视点改变的效果。

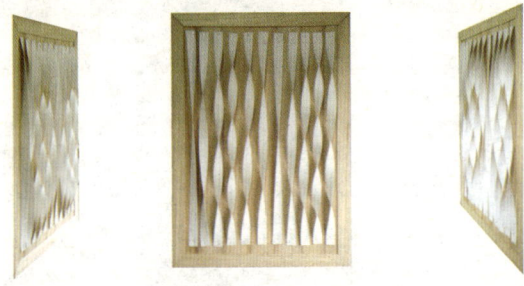

柔软转动的条带形成变化的表面,随观察者视点的变化而变化

作业 2 "建筑是运动的框景"

建筑是相当固定的实体,它承载了人的活动。人的运动是生生不息的,同一个人一天中在建筑中的运动是丰富的,建筑框景了他的活动。而一群人的运动则让建筑充满了生机,展现了人群的生命力。建筑这个载体框景了千家万户的活动。作品创造了众多不同家庭的活动,有孤单老人独自看电视的场景;有大家庭晚餐聚会的场景;有儿童嬉闹的场景……建筑阅尽了人生百态。

不同的颜色,不同的空间

1 "Nothing is deeper than a surface."
Imagine a fixed content on a picture, the audience feeling a different scenery during his observing. How to achieve? As person's viewpoint will change the phenomenon, when people walk around the painting because of different viewpoints, the scenery are also different. Works with white strip rotation to form the orderly images, rotating handle strengthened the changable viewpoint.

2 "Architecture is the box scene of movement."
As a quite fixed entity, architecture illuminates human activity. A person's movement is perennial. The same person in a day has rich motions in a construction, which enframes his activities. But a group of people's movement allows building full of vigor and shows the crowd vitality. Building as carriers enframe thousands of families' activities. Works created many different families' activities. A lonely old man alone watching TV scene, a big family dinner party scene and a children's rowdy scene... Buildings see various aspects of life.

案例 11

缪丹

罗丹说过：所有面积，好像是正在它后面推动的体积的最外露一面。要设想形象正迎着你们，向你们突出。一切空间的力量都在厚度的概念中。一片叶尖向我们伸展的树叶比一片平面的叶子更有生机，我的作品就是从厚度的分割和剖切开始的。

平面实际是三维空间的无限压缩，所以我们看到的平面是空间的一端。当我们取被压缩空间中的若干个截面，并将其拉开一段距离时，再看到的也许会是一个不一样的由平面构成的三维效果。所以最终作品在不同角度看会有不同的感觉。

平面是由无数线条组成的。选择其中一些线条割开，并做三维变换，会得到平面与线条不同的关系和效果。

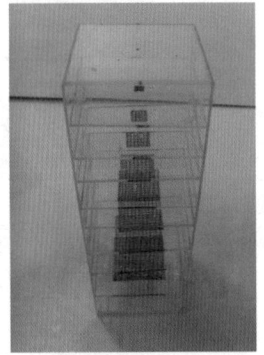

不同角度会有不同的感觉

Rodin said, All the area behind it seems to be driven by volume is the most exposed side. To imagine the image is facing you, to your outstanding. All the space forces embedded in the concept of the thickness. An extension of the leaf tip to us is more alive than the leaves of a plane. My work starts from the segmentation and dissection of the thickness.

Plane is actually a infinite compression of three-dimensional space, so we see the plane is the spatial end. When we take several sections of the compressed space and separate them, what we see might be a three-dimensional effect formed by different planes. So when we see the final work from a different point, we will have different feeling .

Plane is composed of numerous lines. Cut some lines, and do three-dimensional transformation, we will get different relation and effects of plane and line.

在平面上稍作变化，便可得到凹凸质感的树叶

 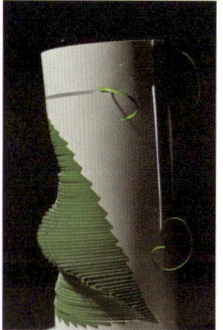

在空间上做变化，便可得到完全不一样的三维效果

案例 12

唐晓兰

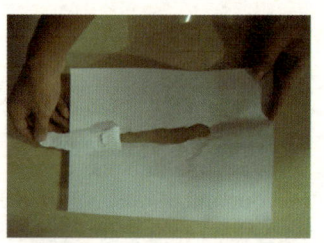

作业 1

关键词 1—突破

"撕裂"这个动作,可以理解为仅仅是撕裂一张纸而已,但是我把它理解为一种突破,这使得平面有了改变,有了三维的意向。每个人会有自己不同的想法,但是相同的一点是我们都认为平面有所不同了。

关键词 2—内和外

正因为有上面的一个动作,就产生了"内"与"外"的关系,也就产生了三个层次—平面内、平面上和平面外,且三者间有所交流。如果我们可以进入到平面的内部,或其内部的东西可以跳出来展现在我们眼前,这又会产生怎样的改变呢?

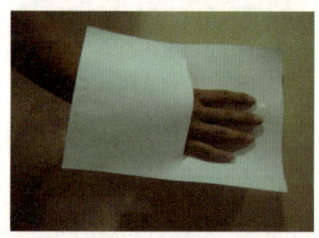

关键词 3—对比

绿色,是生机,代表欣欣向荣;红色,是落红,预示凋零枯败。两者又是互补色,有视觉比较性,但是这不仅仅是简单的对比,而是说明简单的两种色彩,也可以诉说很多内容。

成果

是表面内外的交流、对比和相互依存;

是平面与立体的冲突;

是繁荣后必将萧瑟的定论;

是红与绿的色彩表达;

"没有什么东西比一个平面更有深度",这不仅仅表现在空间的深度上,也表现在这个平面表达的内容的深度上,你聆听到了什么,又有何感受呢?

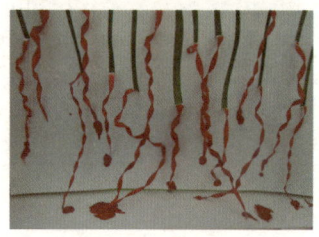

Key word 1 — Breakthrough
 The movement "tearing" can be just understood as tearing a piece of paper, but in my view this act represents a breakthrough which changes a plane with a three-dimensional trend. Everyone has his own understanding, but what we all agree is that the plane is different.

Key word 2 — In & Out
 "Tearing" has created the relationship between in & out and three levels — "in the plane", "on the plane" and "outside the plane" which are interlinked. If we can go into a plane or just in the opposite case, will anything change?

Key word 3 — Comparison
 Green means vigor and prosperity; red means fallen flowers and withering. These two are complementary and comparative colors. Though they are two simple colors, they imply much more.

Final work
 It's the communication, comparison and mutual-relience between in & out. It's the conflict between plane and solid. It's the conclusion that recession surely comes after prosperity. "Nothing is deeper than a plane." is shown on both the depth of a space and the depth of a content. What do you hear and see?

作业 2

 "建筑是对运动的取景。"
 运动，总是伴随着时间的流逝，可以是一瞬间，也可以是一段历史长河；取景，就是对某一个时间点的定格。
 每一条，都表达的是一个事物的发展过程——从胚胎到老年，从种子到森林，从雨滴到大海，从部落到都市，不一而足；每一条在穿过时间的平面时都是一个时间点的定格；
 每一条都是从平面后的绿色渐变到平面前的红色，是一种对比。

 "Architecture is the enframed scenery of movement."
 In my view, movement is always accompanied by the lapse of time, which can be just a shot moment or a long period of time, and framing is the still of time.
 Each strip is the developing process of a object — from embryo to the elder, from seed to forest, from water drop to sea, from tribe to city and so on. It's the still of time while each trip is going through the plane.

作业 3

当线条不再是平面中二维的图形，而是冲破平面，成为三维的、甚至是与展厅空间发生空间联系的物体，会给人完全不同的视觉体验和思考。我希望观者能够有自己的体会。

观众和作品之间到底是一种怎样的微妙的关系？是线条从作品中流动出来？还是观众进入作品去观赏？我希望观者能有自己的答案。

When lines are not just the figure in the two-dimensional plane, but the objects which are there-dimensional, and even connected with the exhibition hall, they will give us brand-new visual experience and ponder. I hope you can attain your own perception.

What is the relationship between an artwork and its audience? Will the lines flow out, or will the audience go into the artwork? I also hope you can find you own answer.

案例 13

田梦晓

本作品重点表达中西方绘画的异同。从"人物"、"景物"、"风景"、"场景"四个角度选取中西方部分绘画作品，分别重叠于四个人面上。通过墨色的交融体现中西方绘画的区别和相似之处。

操作过程中，将选取的画作铺于人像面具上。接着将水泼在纸上。油彩随着水流淌并混合，最后停留在未知的地方。纸也贴合在人像上形成隐约的五官。

The emphases of my works are the similarities and differences existing in eastern and western paintings. From four perspectives of "people","scenery","landscape", and "scene", I select some distinguished works coverd on 4 marks. Embody my emphases through the mixture of ink.

In the process of design, I put the selected copies of paintings on a facial masks. Then, I poured some water on the papers, letting the colors permeate freely,mix up, and finally stay at a certain place. As for the layers of paper, they got wet an soft. Following the shape under them, the paper changed into colored masks after dried again.

案例 14

岳碧岑

人是感性与理性的结合。人用理性去分析客观世界。变化着的各种透视关系是人对世界的认知。人用感性,去感受世界。事物,人物,重叠在一起的,是人对于各种纷繁事物交织在一起的主观认知。

(1) 中心的方框是中性的,而加入了阴影的方框在空间中有了深度。

透视是人对于客观世界分析的产物,而变化着的透视则表现了人对于世界的不同的分析。

变化着的透视关系,给中性的框赋予了不同的意义,也使整个画面更有深度。这深度不仅是画面本身,更是观者与画,作者与画,甚至是作者与观者的思想的深度沟通。

(2) 拼贴是一种记事方式。它可以像电影一样记录这一系列或相关或毫无关系的事件。一张简单的作品,如同框住了一段穿越时空的人事。有的可以让人联想,有的令人思考,或仅仅,只是为了表达一种让观者莞尔一笑的幽默。同样的,拼贴不仅仅是简单的剪切粘贴,时间空间赋予了它思想上的深度。同时,线条、色彩、构图的美感,也同样表现在同一幅图中,每一个画面所承载的,远远大于他表面的价值。

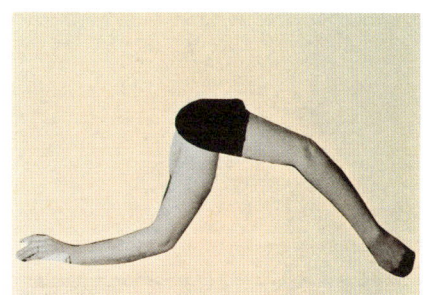  

Human being is the combination of emotion and ration. People use rational part to analysis the objective world, and people use the emotion to feel about this world.

1. The box in the center is neutral while the space has the depth when added the shadow to the box.
 People can imagine what kind of things can make that shadows, and at that time, the visitors and me, the creator, are connected through that thinking.

2. Collage is a way to record.
 A piece of work contains lots of things and it's more like a movie which contains history, imagination and thinking. Sometimes, it shows about feelings or thoughts, but sometimes it is just some expression about humor.

案例 15

张润泽

序言　存在的探寻

现代艺术自塞尚之后从客观描绘世界转入主观意识浓厚的个人创作。这种属性自毕加索、马蒂斯、蒙德里安后达到高潮。从哲学的角度来说，这些艺术家通过描绘自身理性的情绪体验而达到了"存在"。萨特说，存在先于本质。而且存在是偶然的，荒诞的。那么存在该怎么定义？又怎么表现？存在的意义又是什么？

No.1　对话

在一个进深感强烈的空间中，一个人独自矗立。对面存在的是自己的倒影？是粗线条勾勒的虚幻？还是空白的存在？本体和本体的异化体之间的对话，也就是"存在"和"存在？"的对话。这个作品试图探讨并呈现这种可能性，并对存在的本质提出质疑。

No.2 燃

这是一副湿画法画成的水彩画。

颜料在水中四溢,

在纸上留下自然的水痕。

然后,我试图用火焰来寻找一种形式上的对应。

在展出现场,我把这幅悬空的画作烧掉,不断向上跃起的火焰和向下流的水痕形成对比,展示时的片刻欢腾和展后的冷落凄清形成对比,同样,展示前的完整画作和展后的灰烬形成更惊心动魄的对比。相信作品也是有生命的,它的存在究竟是展示前的完整?还是展示时的疯狂?抑或化为灰烬后的冷清?

整个过程既是我的作品。

它宛如一次凤凰涅槃,

在灰烬中重生的是一种意义。

它激烈的展示了存在与毁灭的过程,

启发人们去思考深层的存在的意义。

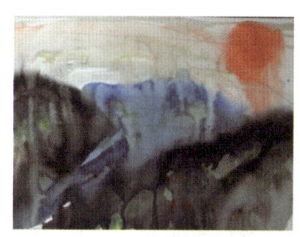  

结语 存在的意义

老子说:
大音希声，大象无形，大方无隅，大器晚成。
在我看来，这句话意味着:
短暂的存在必将给人留下长远的回味，
空白的存在必将给人以无限的遐想。
之于广阔无垠的浩瀚时空，
每个人都是沧海一粟，享有白驹过隙的一瞬光阴，
如果我们在这一瞬看透了表象，燃烧了自我，
即会得到生命的涅槃。
也许这就是存在的意义。

Preface Inquiry for existence

Modern artists turned into subjectively presenting the world from objectively depicting it since Cezanne. And this trend came to its acme after Matisse, Mondrian, and Picaso. From the philosophical point of view, these artists achieve their condition of "existing" through presenting their own emotional experience. Sartre said, existence precedes essence, and existence is fortuitous and absurd. Then how do we define existence? How to present it? What is the signification of existence?

No.1 Conversation

In the space of definite depth, a man stands alone. Is the existence on the other side the shadow of himself, illusion of wide streaks, or nothing? The conversation between thing itself and thing`s mutation is the conversation between "existence" and "existence?". The work tries to probe and present this possibility and doubt about the signification of existence.

No.2 Burning

This is a watercolor drawing. Pigment flow with the water, and leave the trace naturally. Then, I try to find a kind of contrast using fire. I ignite the drawing in the show. Fire rushing up contrast sharply to the trace of water flowing down. Temporary commotion of the showing contrast sharply to the enduring tranquility. Meanwhile, the intact drawing before the show contrast significantly to the residue after showing. I still believe that works have their own lives. Is its existence the intact being before the show, or burning crazily in the show, or the tranquility of the residue?

My work lies in the procedure of the show. It is just like the burning of Phoenix Nirvana. What reborned in the residue is kind of meaning. It violently showed a course of existence and destruction, inspiring people to think about the meaning of existence.

Epilogue The significance of existence

Laozi said a famous words. But i guess this quotation means that existence of a moment could make people to think over and over, and existence of the blank could make people to think more and more. Each and every one of us is just a pebble of a large mountain, possessing a jiff of the limitless history. Had we look through the surface of truth and burning ourselves, we would reborn. Maybe that is the significance of existence.

案例 16

郭欣欣 黄卿云

一张白纸，如何正反折叠着，就构成了造型各异的三维物体？这是一个有趣的经验，让很多理性的科学家为之数学分析，甚至编出程序，计算折痕。在展开的纸上留下的折痕，清晰地展示着从二维到三维的转换过程，为平面与空间的关系提供了新的解读方式。

折纸的纹路深浅变化显示着折纸作品的过程，同时又暗示着最终作品的形态，从平面中感知立体。

How could a piece of paper makes various 3-dimensional objects by folding and folding? It is an interesting experiment. Many scientists even search in math-analysis or making computer programs to calculate the folding trails. It is the trail that leads us to know the changes between 2-dimension and 3-dimension.

Through the tiny lines of the paper folding, people can imagine the shapes they used to be. People can perceive the 3-dimension in 2-dimension.

案例 17

路天

"没有什么比一个表面更具有深度。"

关键词：无限、映像、透明

1.无限
拥有进深感空间，给人以不断延伸、无穷无尽之感。

2.映像
水与光都能使物体产生与之对应的相似的映像。物的原像与映像层叠交错，是视觉上的真实与虚幻。

3.透明
透明性产生多重空间，物质与空间的透明体现了浅空间与深空间之关系。

装置说明
第一层是白纸，其余每一层纸片都打有小孔。将一定数量的纸片重叠。
N个简单的平面相加=一种富有深度的空间感受。

光从装置的后侧照入，观察者从观察筒观察。通过光源上下、左右、前后的移动，以及移动的轨迹、频率、光的颜色不同，产生不同的效果。

最终成果　穹

设计说明：通过营造一个半封闭的空间，在空间上方安置作品。作品整体为白色，白色是简单、纯净的色彩，给人毫无瑕疵的感觉。当观众在幽暗的半封闭空间仰头观看，犹如置身在无边无际的冥想空间。平面给予人无尽的想象空间，就像是仰望星空时的感受。

装置说明
作品为30cm×30cm×30cm的白色立方体，在立方体中部放置数层白纸，第一层是白纸，其余每一层纸片都打有小孔。同时在装置上方安装25W的白炽灯作为光源。

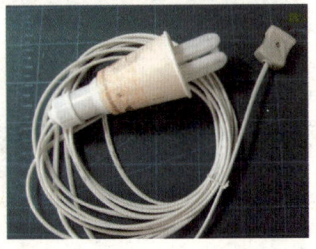

The Night Sky

Inspiration origin: Perhaps many people are like me, who like looking up into the night sky and exploring the little twinkling stars which hide behind the dark blue curtain. Watching into the twinkling night sky, I seem to hear the songs gently from a grandmother of a child, and the childhood days playing with the junior partners seem to be back. It seems, floating to fly up and turns into a bright shining star in the night sky. Whether near as the moon or far as the planets tens of thousands of light years away, are pieces of finished project to our eyes, so that you cannot tell how far they are, and cannot see through the deep night sky. When I was looking up into the night sky, I feel like being in an eternal space, an immense meditation space.

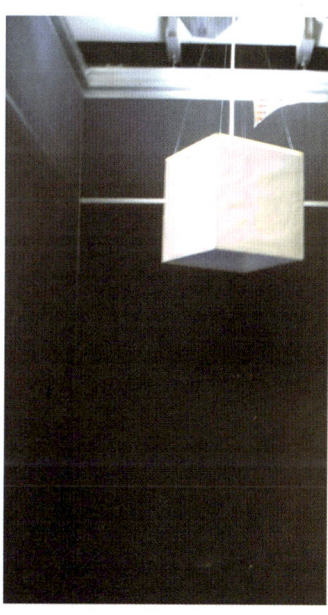

Design explanation: The main device is a 30cm×30cm ×30cm white cube with several layers of white papers. In addition to the first layer, the other layers of papers have each punched a few of small holes in the same size. The device is installed with a 25W incandescent lamp on the top as the light source. Observers can observe the work from the observation tube. By gently shaking the device, the relative movements occur around the device and the light. The device is white as a whole and white is a simple, pure color, which gives a feeling of flawless. The device placed in a semi-enclosed space, when audiences watch the work in the dark, they will feel as if they were in the middle of the endless meditation space. In short, even as a surface, through the lighting, gives people the endless imagination, like the feeling when one was looking up at the stars into the night sky.

案例 18

张栩然 何雅 吴晓庆

形式：装置整体呈线形，流动的液体象征时间的流逝，自上而下，简单的几何形体和四色的彩带暗示着时间匆匆而过，却能留下色彩斑斓的轨迹。

灵感理念：时间是指尖流沙，难以捉摸，记忆却可以染上颜色。似水流年，春夏秋冬，逝去的是时间，沉淀下来的是碎片般的点点记忆。一天天，一月月，一年年，记忆就是一点一滴的时间渐渐积累，慢慢沉淀，不知不觉中看似一点一滴缓缓流过的无数个日子碎落成丝丝记忆，模糊又清晰在脑海里。

技法：作品通过在针管中注入墨水，使其在重力作用下顺着管道流入宣纸中，慢慢渗开。日历通过象征生活的碎纸机变成碎屑，再注入象征四季的玻璃盒中，象征四季的汇总。

Form: The overall linear device, a symbol of the flow of the liquid passage of time, top-down and simple geometry with four-color ribbons implies time goes by, but are able to leave the colorful track.

Idea: Time is like the quicksand between fingertips and elusive, but the memory can be colored. With years rolling like a stream and the changing of four seasons, time goes by but the fractions of memory precipitates. Day by day, month after month, year after year, memory is gradually accumulating little by little. Slowly precipitating numerous days flow through unconsciously. At last, memory break into pieces, vague and clear in my mind.

Techique: Inject the ink in the needle tubing so that it flows under gravity down the pipeline in to rice paper, and slowly infiltrating. Calendar are shredded through a shredder, a symbol of life, into debris, and then are injected into, a symbol of the four seasons, the glass box which is a symbol of a summary of the four seasons.

案例 19

吴超楠 姚远

1. 没有什么比一个表面更有深度

在此，我们希望探讨的是二维和三维的转换。在构成主义代表人物马列维奇的作品中，他以三维表现了二维。在杜尚的绘画作品中，他以一个简单的动作突破了平面的表现。我认为二三维之间的转化是简单的，我想探讨的是，二三维的关系的问题。

于是产生了一个简单的设想，我们观画无论以什么视角，都是在三维的世界观察二维，而假使我们处在一个平面世界中，可以看到的三维世界会是怎么样的？会是像中国古代的散点透视画，或者毕加索的立体主义？透过一个平面，我们世界变成了画面，画面本身转变成了一个实质的世界，我们成了画的内容，而画面内部的手是创造我们的画师。在此，表面的深度被发掘到了无限的层次。

Nothing is deeper than a space.

What we are supposed to discuss about is the transformation between the two-dimensional and three-dimensional space. In the works of Kazimir Malevich — a typical representative of constructivism, he showed two dimensions in the form of three dimensions. And in the drawing works of Dusan, he made a breakthrough on the expression of plane only by a simple action. In my opinion, the transformation between two and three dimensions is easy and what I want to discuss here is the relationship between the two.

A simple hypothesis occurs to me. When we observe a drawing, we are observing a two-dimensional object in a three-dimensional world, no matter how we change the angle we look at it. But just imagine that the world is two-dimensional, then what will the three-dimensional world be? A drawing of cavalier perspective in ancient China, or the cubism of Picasso? Our world can be turned into pictures and then be turned into a substantial world all by a flat surface. At the same time, we have became the content of the drawing and the hands inside are the creators of us. The depth of the surface can be perfectly displayed in this occasion.

2.线条的图像 图像的线条

在此,延续了上一个主题,依然是探讨转化的问题。依据霍金的理论,低维度之所以呈现出低维度的特征是因为我们没有能足够精确的放大。譬如一个蚂蚁对于一片树叶它是一个面,而对于地球,它就是一个点。当我们拉近到无限近,它依然是以体的形式出现的。

假设我们把地球的一条纬线展开,它能呈现出面的特征,像一幅风俗画,但同时它又是线性的。诸如此类的关系还有很多,不再赘述。最后,我们想到了借用《清明上河图》这样一个明显的线性的中国画表现对此的思考。

在无限远的表面它以粗线条的形式呈现,而借助一个足够精确的放大镜,我们实现了人眼分辨率下图面的展开,从而证明线与图像是可以转化的。

The picture of lines, the lines of pictures.

We are going to continue the last topic: the question of transformation. According to the theory of Hawking's, the reason why we can characterize the low dimensionality is that we are not able to magnify in a large enough accuracy. For example, comparing to a leaf, an ant would be a plane, but comparing to the earth, an ant would be a point. When we are forced to stay very close to observe all this, it will be displayed in the form of the body.

Assume that we can unfold one of the earth's parallel and make it into one flat surface so that it can show its characteristics of a plane and it will be like a linear genre painting. There are a lot of examples for this. At last, we want to express our thoughts on this problem by a traditional Chinese painting—"Riverside Scene at Qingming Festival", which is a typical linear painting in China.

It has proved to us that a line and a picture can be transformed to each other because its lines at the infinity point are presented as thick lines, which become gradually clear enough by means of a magnification with large enough accuracy.

3.表面与空间的互动

在此，转化不再是主题。我认为事物的关系探讨应该包括转化与互动两方面，既然低维度和高维度之间是能够相互影响的，那么我认为可以用一种渠道将两者以直观的形式联系起来。

安格尔的名画《泉》中最主要的入画点我觉得是那个瓶口，出画点是那个脚底下的泉。对入画点，进行重新的演绎。假设那就是一个媒介，从无限深度的画面后，用光的形式代替水流出来，到三维世界的杯子中。这样关系就升华到了互动的和谐。我们顺着入画点会发现回到了一个真实的物体，于是，出画点就成了我们这个世界本身。到底哪里是画面，哪里是真实？或许语义上已经不再清晰了。

The interaction between surface and space.

In this chapter, transformation is not the theme. I believe that the relationship of two objects include two parts. One is transformation, which we have already discussed. And the other one is interaction. Since there are mutual influence on each other of low dimensions and high dimensions, I think there exits a direct link between them.

In the famous painting "Spring"(by Engle), the two significant points are the bottle neck and the spring at the foot. We can give a illation on the first point. Assume that is a medium, and the things will come out in the form of light instead of water flowing from the great infinity of the picture, coming to the cups in the three-dimensional world. In this way, the relationship will sublimate to a very harmonious state of the interaction. We can know some real objects when we observe the drawings carefully. In the end, we are back to the world itself. Maybe we will be confused and lose awareness of the differences between the picture and the reality and words may fail us in expressing our feelings.

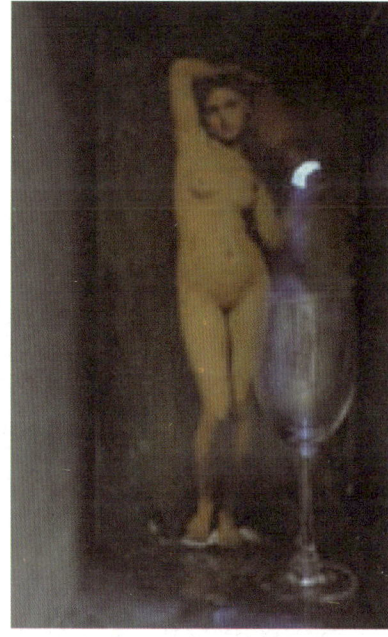

案例 20

李欣路 王里漾

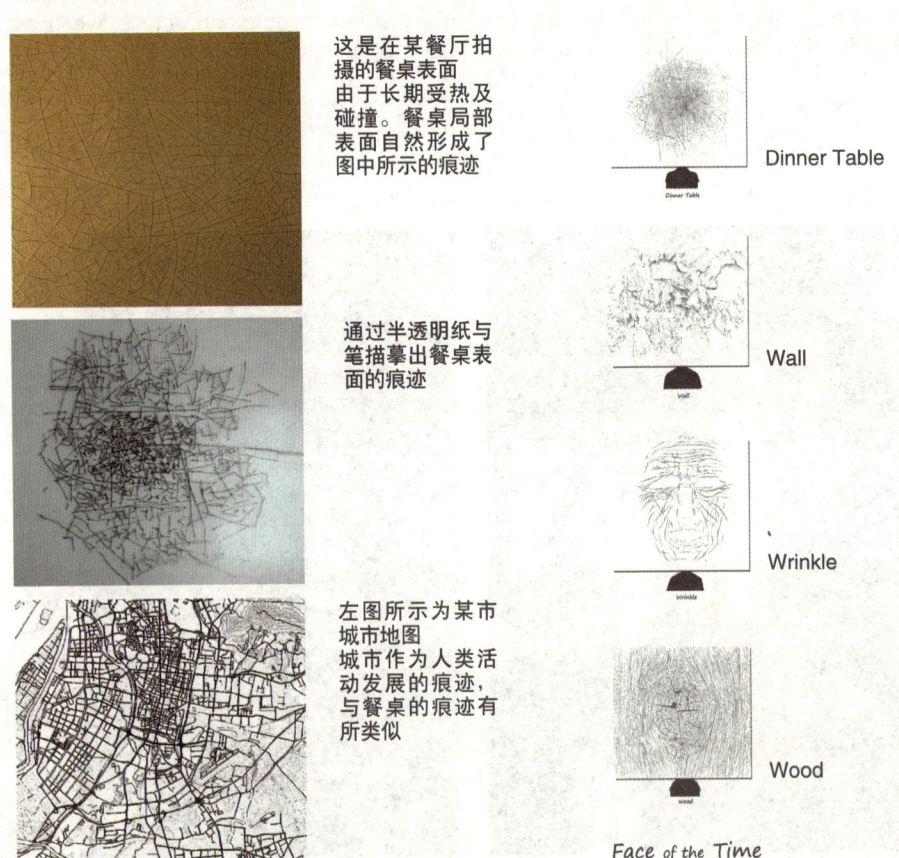

这是在某餐厅拍摄的餐桌表面 由于长期受热及碰撞。餐桌局部表面自然形成了图中所示的痕迹

通过半透明纸与笔描摹出餐桌表面的痕迹

左图所示为某市城市地图 城市作为人类活动发展的痕迹，与餐桌的痕迹有所类似

Dinner Table

Wall

Wrinkle

Wood

Face of the Time

线条的韵律

在餐桌痕迹的基础上,我们继续搜寻了一些线形丰富的物体,将其反转颜色进行描绘,企图寻找出线条的韵律

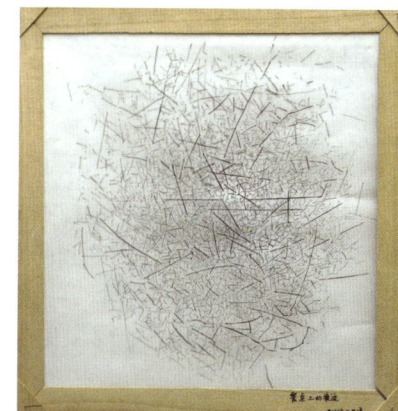

用任何东西在展出的石膏中做出的痕迹都是崭新的时间的脸

案例 21

于善君

一系列的作品以文艺复兴时期开始出现的透视法为出发点，尝试将立体和平面上的图形相互投射，在平面与三维物体上呈现不同的视觉效果。

作业1

平面与立体
透视法
投影

文艺复兴时期，开始出现运用透视法绘制的画作，画家将感知到的物象转化成平面与平面的关系，即将三维场景投影到画纸平面上，令视像更具真实感，更详尽表达物体的存在状态。

同样，将平面的图形投影到三维物体上，也可削弱三维物体的凹凸感。

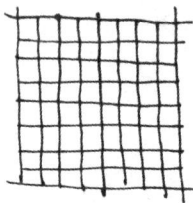
平面的网格

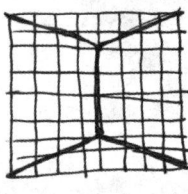
将平面正交网格投影到三维物体上

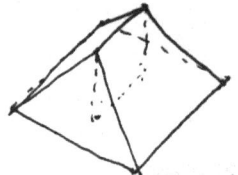
底面为正方形的突起物体

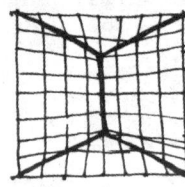
将凸起物体上的正交网格投影到平面上

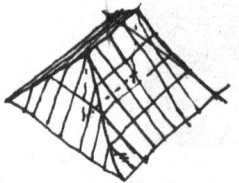
上图物体加以正交网格

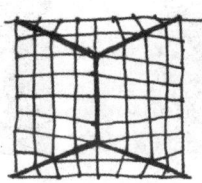
将下凹物体上的正交网格投影到平面上

以黑色为底色，尽可能减小阴影明暗对视觉的影响，从正前方可以观察到具有深度的平面图形和平面化的三维物体。

From the Renaissance period, start to appear use of perspective drawing paintings, artists will perceive things into plane and planar relation, upcoming 3d scene onto the plane, another video paper, more realistic in greater detail express object state of being.

Similarly, will planar graphics onto the three-dimensional objects, also can weaken the three-dimensional objects concave-convex feeling greatly.

With black base, minimize the shadow of visual impact, light and shade from the front can be observed depth of the plane figure and the complanation three-dimensional object.

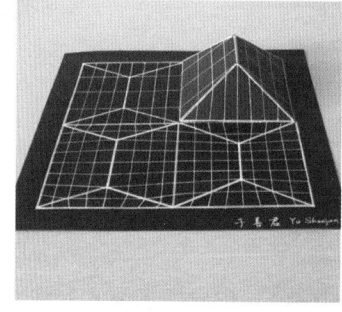

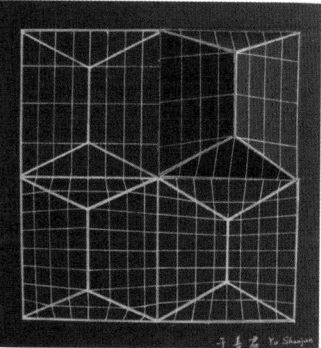

作业 2

平面
倾斜面
笛卡尔坐标系空间
索尔·勒维特（Sol Lewitt）

20世纪60年代倡导概念艺术和极简主义的标杆式人物，仅用最基本的四边形、三角形和圆形，以及最简单的三原色和黑色，反映外化为几何造型的思考内容。

如果将平面也视为空间，分别以黑白网格和三原色色块表现的两组三种不同深度的空间在某一点观察时可以获得同样的视觉效果。

In the 1960s, the benchmarking type character advocating concept art and minimalism use only the most basic quadrilateral, triangles and round, and the most simple colors and black to reflect the thinking content in the form of geometrical formative .

If the plane is also regarded as space, Observing two groups of three spaces of different depth expressed respectively with black and white grid and tricolor color piece at a particular point can gain the same visual effect.

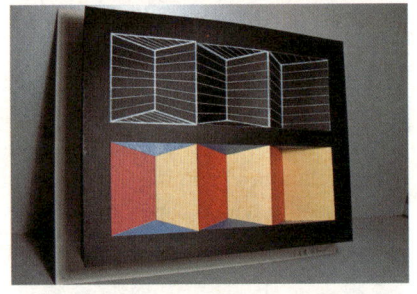

作业 3

舞台的深度

奥林匹克剧院-帕拉迪奥，1508-1580

剧院的布景由几条向后不断抬高变窄的巷道组成。在两侧及地面上有按照透视法绘制的壁画。前面的高度连续后退，由突出的雕像从视觉上来补偿；借助突出部分和壁龛的合理组合来增加深度的想象力。

将现实空间、倾斜面和平面的图形相互组合，以黑白网格和色块分别表现抽象的舞台空间。

将现实空间中垂直面的上下交角分别替换为平面图形和倾斜面，在较浅的空间中创造现实空间应有的深度感。

The theatre sets for a few backward continuously raise effacement of roadway composition in both and the ground was covered in accordance with the perspective drawing murals. In front of the highly continuous retreat, by outstanding statue from visual up compensation, With prominent part and niche reasonable combination of the increase of the depth of imagination.

Will the reality space, QingXieMian peace graphics mutual combination, with black and white grid and color piece of abstract stage respectively performance space.

Connect reality space perpendicular planes fluctuation JiaoJiao respectively to replace the plane figure and QingXieMian in shallow space creates reality space should be feeling of depth.

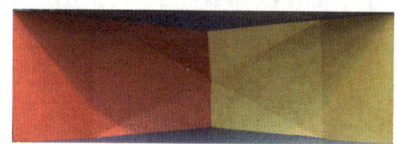

案例 22

张蓝兮

设计由时间起源，发展出人生的概念，以女人的一生为载体，采用简洁的线条勾勒出女人在生命各阶段的轮廓，表现生命的运动。当女人渐渐老去，发丝干枯，皮肤苍老，女儿却大了。看到她仿佛看到了自己。这是生命的伟大轮回。设计过程参考了克里姆特的《女人的三个阶段》和杜尚的《下楼梯的裸女》。

The design, originating from the idea of time, developing the conception of life, taking a woman's life as the carrier, and sketching a woman's diffrent life stages by concise lines, shows the movement of life. As a woman got old, with the shrivelled hair and the aged skin, the daughter grew up. It is like seeing herself when seeing her daughter. It is life's great samsara. I consulted the "Three Periods of Women" of Klimt and "Naked Women down the Stairs" of Marcel Duchamp.

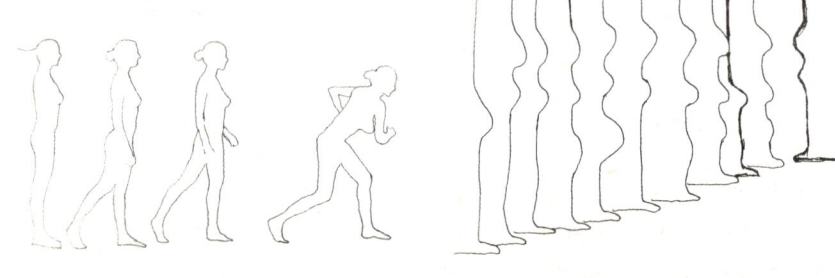

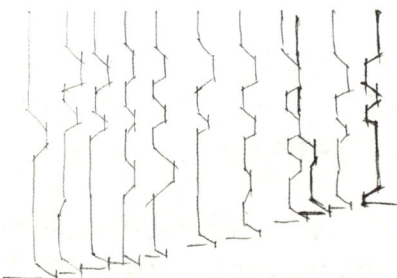

第二个习作仍然延续了习作一的探讨人生的主题，采用一半人脸一半骷髅的形象，意在表达"体味生活，淡定生死，活在当下"的禅意主题。设计以黄白黑条纹相间这样一种拍电影时常用到的标志的形式，表达了一丝"人生如戏"的意味。本次习作以线条为表现手法，通过线条的弯曲排列在平面上表现出立体的效果，与主题"图像的线条，线条的图像"相对应。

The second project continues the topic of the first project, using the image of half face and half a skeleton, to express the zen theme of appreciation of life, being calm to life and death and living in the present moment. Design implies the idea of "life is like a theatre" with white, yellow and black stripes as a sign of form that films often use. In this project, lines of bending arranged in planar express three-dimensional effect. And the design corresponds with the theme "image line, line images".

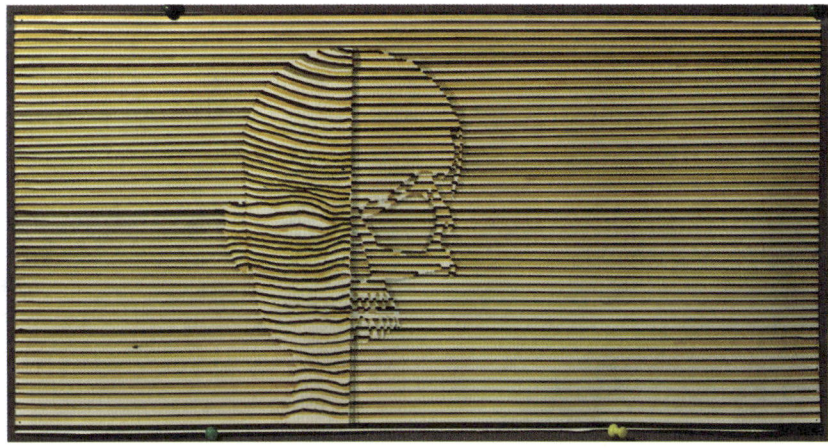

习作三在表达主题上与之前的习作一脉相承，仍然是对女人、人生问题的探讨，在此不做赘述。

采用两个平面拼接的手法，形成一种视觉上的三维空间的延续的错觉，同时使用脚印这样的符号来体现时间流逝的主题。

加入达利的柔软时钟的元素，进一步深化时间的主题。

The third project still continues the topic of the last two — the discussion about the theme of life and women, so we do not talk about with details.

案例 23

李丹 李林希

最初的想法是通过简单的元素排列、组合和变换产生一种独特的韵律和感受。

作品通过木条的排列和黏贴，表达意象。木条紧密结合的部分，是一个脚印的意象。而周边的部分则以间隔排列的木条表达。整体做成一个倾斜的平面，而比较实的脚印部分则倾斜地陷入其中。为了表达随着时间的流逝，我们将周边的木条排列由密集逐渐变得稀疏，到最后，只有星星点点地出现；而实心的脚印形象则完整地出现在整个作品中，从而被凸显出来。

整个作品希望表达的是一种对行走中，消逝的和留下的的思考。我们行走于人生之路上，经过的很多东西、很多事情都像浮云一般，消逝无声。然而，有些东西，却深深地留存下来，就如同作品中的脚印。别的东西越腐朽，就越凸显其珍贵。那些消逝的，成为记忆；而那些留存的，成为印迹。

The original idea was that the arrangement, combination and transformation of the simple elements generate a unique change of the rhythm and feeling.

We experss the images by the arrangement of battens. The closely integrated part of the battens is the image of a footprint.And Surrounding area is made by spaced piece of battens. We make the whole to be a sloping plane. And the solid part of the footprint tilts into it.To express the passage of time, we arranged the battens to be more and more sparse, till the last sprinkles.What's more, the image solid footprint appears completely in the whole work, thus highlighted.

The work hopes to express a thought about the relationship between the disappearing things and surviving things during our life. We walk on the road of life, experiencing many things. Some just like clouds go away silently. However, some things, just as the footprint in the work, deeply retained. Something else decadent highlight its preciousness. Those who passed become memory and those who retained become imprinting.

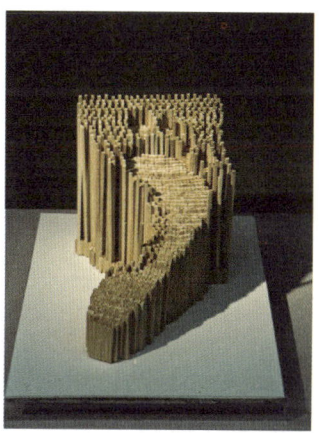

案例 24

丁立南 许碧宇

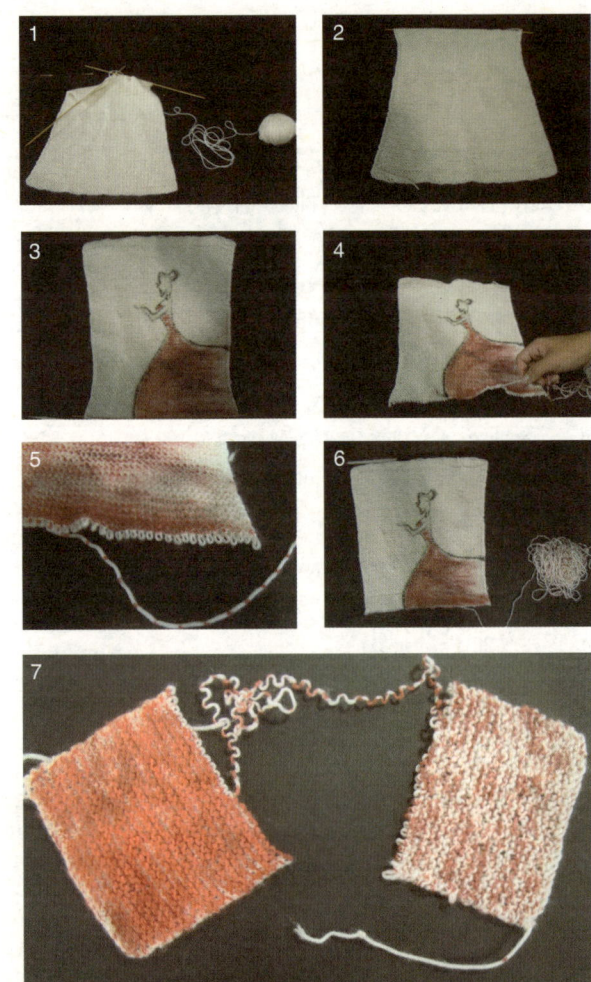

　　《编了又编》这个作品的主题是"图像上的线，线上的图像"。我们试图寻找图像与线重组与转化的新途径。我们借助编织这一传统手工技艺，首先将线状的毛线编织成面状的毛线片，即将线组成一个面，可将其看做一张平面的画布。接下来我们在画布上作画。画作完成后又将画布进行拆解，使之重新变成线。但这些线上已经有了画布上的图画。最后将拆下来的线又重新编织使带有"旧"图像的线重组出有"新"图像的面。

　　后来我们进行了其他类型编织的尝试。比如采用粗麻绳编织绳结，仍像之前一样绘画后进行拆解再编织。

The work aimed to study the relationship between a surface and its depth. There was plenty of ways through which we can understand the depth of a surface. Here we chose two abstract elements: information and time. By using the texts and graphics in magazines and newspapers as the "carriers" of information, and uniting them, we created a surface that can show us its depth of information. Then we used transparent plastic sheets to make the surface separated on the "axis of time" significantly. A letter "e" which was seen as a symbol of information flew off the surface, also on the axis of time.

 The theme of the work called "Weave and Weave Again" was "the lines of graphic and the graphic of lines". We tried to find a new way to make each of the two elements transform into the other. Traditional knitting is employed here. To start with, we knitted a flat piece with yarn to paint on. After painting on the knitted piece, we continued by unweaving the piece to get the yarn with the color we had just painted on. With this, we re-knitted the piece so that we could produce a new piece with the original paint.

 Later we tried other ways of knitting such as knotting with manila ropes, and the same method stated above is adopted.

首先，我们用绳子进行编织，形成条状物。某种意义上来说，这可以看做是由线组成了体。然后，在某表面上涂上蓝色染料。

接下来，我们将染好色的条状物进行拆解，然后用其他的两种编织方法进行再编织。由此，我们得到了不同的表面构成。

案例 25

黄敏 马加伟

根据对"没有什么东西比表面更有深度"这句话的理解，我们分别进行了不同的构思：

（1）"颜色状态"：用不同颜色，不同角度而产生不同人物状态的手段进行解释。

（2）"奇怪的人"：用人物两个差异较大的表情构成一张脸，运用简单的手段进行表达，诠释其解释。

（3）"片中光"：运用灯光的强度渐变说明其道理。

奇怪的人

人是一种奇怪的生物，拥有多面性，人类所特有的情感处于一种很矛盾很复杂的状态。甚至他自己都不清楚她的感情是处于一种什么样的状态，可能是悲伤与开心同时并存的一种状态。这样有时就会处于一种迷茫的状态，就会希望要是能够有一种类似于格栅的东西，让你认识自己，看清自己……

According to the understanding to the "nothing is deeper than surface", we were studied respectively different idea:

(1) "Color status": use different color, different angle and produce different character states to interpret.

(2) "The strange man": use two different expression to constitute a face, use simple means to express, interpret their explanation.

(3) "Light in slice": use light intensity gradient to explain its reason.

The strange man

Man is a strange creature with versatility which is peculiar for mankind. Emotion is quite complex. Sometimes people's emotion in a very antinomy complex state. Even he doesn't know what kind of state his own feelings —maybe a coexistence state of sadness and happiness. So sometimes people may be in a confused state and hope to have something like grid to let him know and see himself...

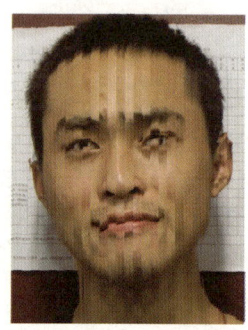

案例 26

闫顺凯 黄潇

　　表面不只是平面，边界之中，它可以由多种元素组成，线条、颜色、光影……因而有空间存在。

　　基于此理解制作的装置可以框出延伸不同方向的空间。

案例 27

胡培

深不见底

有个成语叫做"深不见底",但是如你所见,只有"深不见","底"消失了,不见"底"瓶中的水一直在减少,但是左边的俯视图里,那三个字仿佛无动于衷,始终保持着那个样子。所以如果你只是观察左边的俯视图,你根本不能确定水面有多深。在生活中你看到的三维世界实际上是由一个个平面图像组成的,你怎么能确定这个平面后面有什么。因为就像左边的图像一样,始终是正对着那个表面。

不破不立

用现代手法表现西方传统绘画,象征传统在现代社会中支离破碎,而那静穆的黑色十字以及救赎的白色天光,引导我们完成新时代的艺术手法和精神家园的重建。

There is a idiom "too deep to see the bottom", but you can only see the "too deep to see", "the bottom" is missed. The water in the bottle is losing ,but it is not responsed in the image in the left. It seems like can not feel the change. So if you watch the images in the left, you will not realize how deep under the surface. In our life ,the 3D world is made of many plane ,so how do you make sure what is hide behind the plane. Because just like the left images ,you are up against the plane all the time.

I use the modern ploy to comport the western traditional drawing. The "bottle" indicate the tradition is broken in the morden society. But the black latin cross and the white skylight lead us to accomplish the rebuilding of new art system and spirit homestead.

案例 28

陆明玉

NO.1
在空白平面内的人被限制在二维坐标上，平面不能显示其所具有的深度。

NO.2
具有重复线框的平面显示出纵向深度，平面向三维空间转换，平面中的人被限定在三维空间。

NO.3
将空白的平面延伸至三维空间，被限定的平面即转变为无限的空间，平面即产生无限的深度。

NO.4
将线框的平面延伸至三维空间，产生深度的平面却变为被控制的三维空间。

总结：
不同的视角即产生不同的深度空间。

NO.1
In the blank sheet of paper, we could not find the depth of the place where two men stand.

NO.2
In the sheet of paper which is repeated by black and white frames, we find the two person transform from the plane to the space and we can easily see the depth of the paper.

NO.3
While when we change the coloured paper to a three dimensional expression, we can see the two person are standing in a fixed place and the depth are not so infinite as we see before.

NO.4
As we change the blank paper to three dimension, then we can see the infinite space in it.

No.1

No.2

No.3

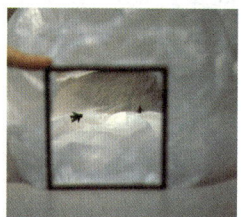

No.4

案例 29
沈禾微 和嗣佳

本设计由"空间中的一点"这个题目出发,结合"建筑"、"运动"、"人体"等因素生成。根据物理中相对运动的经典观点,我们可以说"人在空间中运动,空间在人中运动"。但由于空间是一个不存在的东西,为达到这个目的,我们运用了部分抽象的建筑构件,穿插在人中,为了加强空间穿透性和视觉震撼感,把人解构为一些肢体。这样,所有因素都受到了相对的干

先单纯地用人的肢体互相拼接,可在一张图中表现多个行为。

引入空间……

扰，作品呈现出来的是人与空间的同时变化，在这中间，我们没有严格的参照物，参照物只有我们自己，因此我们还在作品中加入一把椅子，试图将参观者拉入设计之中。

事实上，这个想法最初是模糊的、盲目的、希望以动态表达方式进行，但由于客观条件的限制，我们只能想方设法用静态模型来表现。做建筑空间所表达的是这一过程的后半部分，而建筑师完成的是前半部分，并且对于很大一部分艺术范围似乎都适用。

在这样一种环境中，人和空间达成了一种默契：如果人的活动是有秩序的，那么带来的空间形态也应是有秩序的；如果人的活动是混乱的，那么带来的空间也是混乱的。虽然我们同处在一个空间内，但由于不同的、自我的、唯一的运动，带来的是相对的空间，进而反馈回我们自己的大脑，形成感受。这个空间的固有形态并不重要，重要的是非固有的那一部分，属于我们每个人的那一部分。

运用人的肢体建构空间的一些细节。

  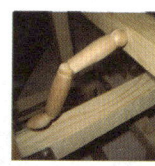  

FROM SURFACE TO SURFACE

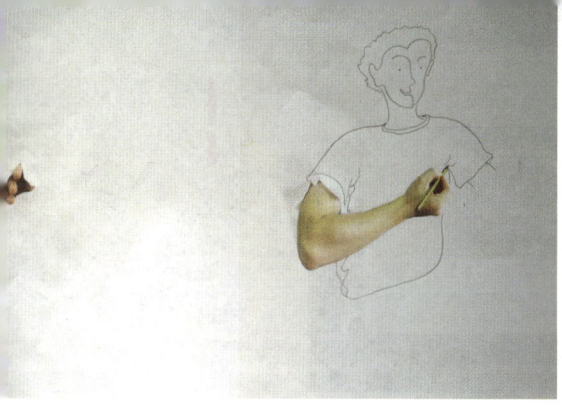

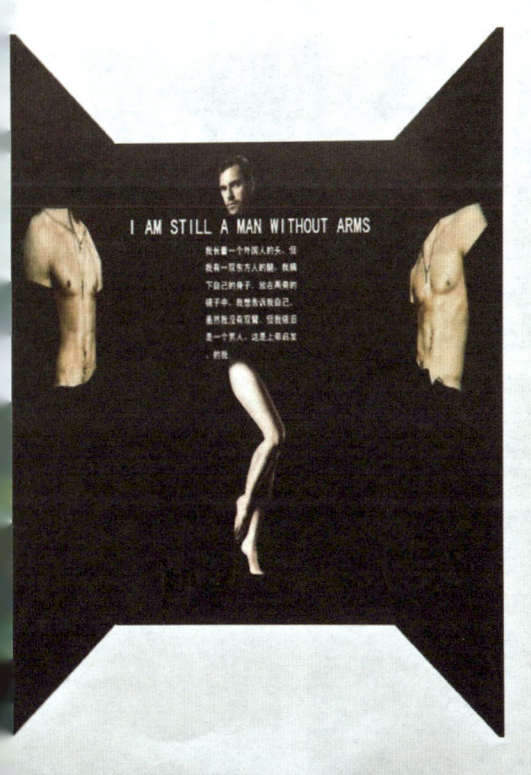

I AM STILL A MAN WITHOUT ARMS

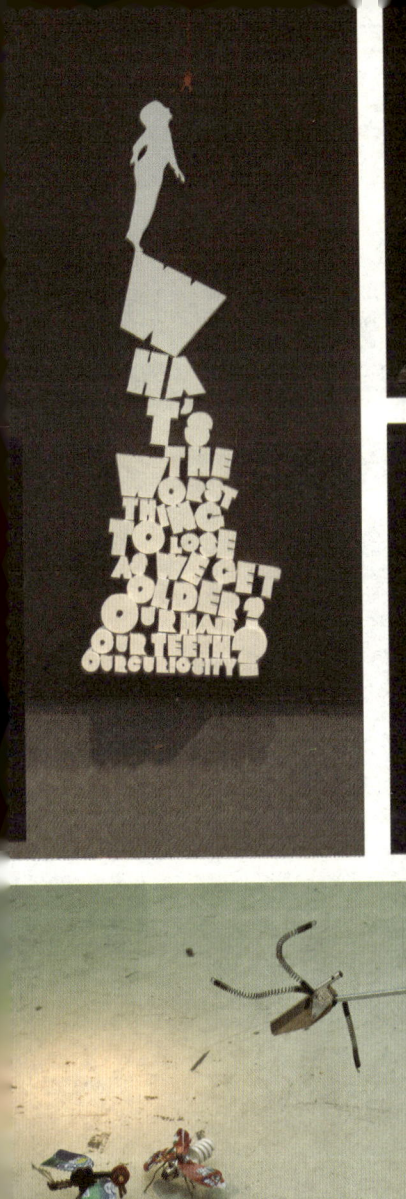

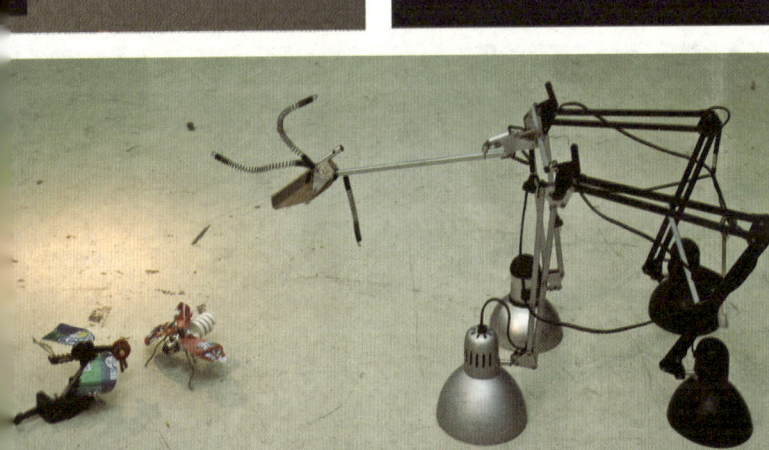

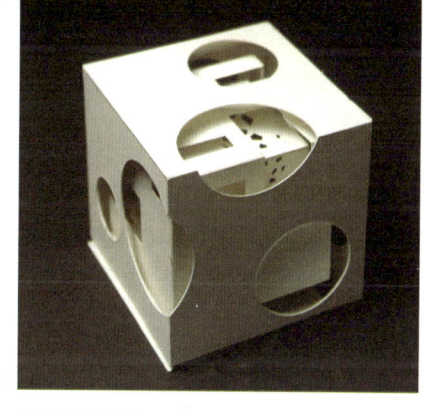

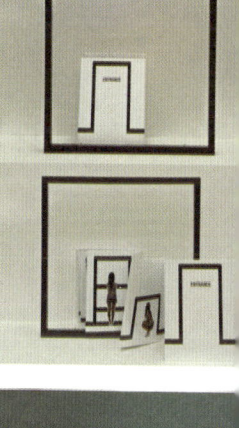

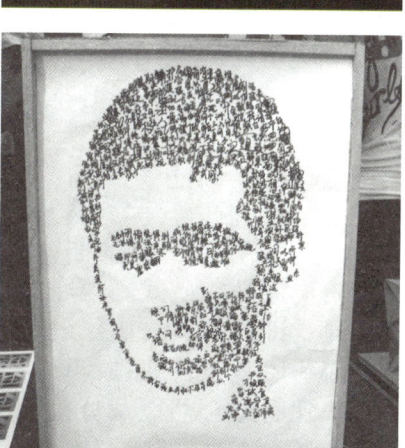
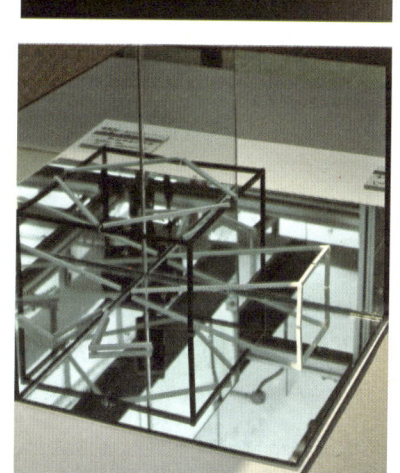

4 课后文集

视觉设计基础教学

曾琼　东南大学建筑学院环境艺术设计系副主任、副教授

联合教学是教学改革深入过程中的重要阶段,有助于我们了解吸收先进的教学理念和方法。早期我们同香港中文大学进行了联合教学的尝试,并请顾大庆教授指导了视觉设计课程的教学实践。这次,在法国国立高等建筑学院葛汉教授的指导下,我们的教师和学生们一同完成了这次联合教学,这是我们在教学改革过程中的又一次实践,也是走向国际化的一次有益尝试。

通过不断的教学改革的实践,我们逐步形成了目前新的教学体系课程——视觉设计基础。"视觉设计基础"课程整合了我们近二十几年的改革成果和经验。它是在吸收了传统的教学体系里经过实践证明了的许多优点后,根据现代建筑设计和艺术设计的发展规律和新的理念,并融合了现代艺术中的各种视觉实验的精华,结合建筑设计和艺术设计的教学特点,整理和发展出的一个全新的教学体系。

教学特点与教学方法

"视觉设计基础"课程是一个综合式的视觉训练模式。这个综合式的视觉训练模式打破了一般训练先简单后复杂、先技能后运用、先基础后综合,以及分门别类的单一教学方法,学生从一开始就要面对比较复杂的视觉和设计问题。结构有序的教学框架,透明化和分解式的练习过程及明确界定的练习目的和方法,这些是此模式的基本特点。

"视觉设计基础"课程由许多专题组成。每一专题都包含了若干个操作步骤,练习的推进和步骤之间的连接都依循"讲课—练习—评图—练习"这样的方法。讲课的目的主要是在一个新的任务的开始简单介绍主要的问题,所涉及的内容只是足够学生开始作业,而不是灌输那些他们尚不能接受的内容。学生在亲身体验的基础上,通过作业的讲评对训练的内容有了进一步的认识。随后的课外作业一方面进一步强化在课内所学的内容,另一方面也是学生独立创作的阶段,检验其对教学内容掌握的程度。

4 Texts and other comments

Fundamentals of Visual Design

Zeng Qiong(Vice Professor, and Vice Director of the Envionmental Art Design Department of the School of Architecture, SEU)

Fundamentals of Visual Design assimilates the merits of the traditional educational system that have withstood the test of time. Based on new concepts of modern architectural and art design as well as their developmental principles, the course fuses quintessential experiments in visual perception from modern art with the teaching styles of architectural and art design to develop an all-new teaching system.

Characteristics and Methods of Teaching

Fundamentals of Visual Design proceeds through a comprehensive course of training in the visual arts. The course breaks apart from the common method of progressing with single, rigidly divided lessons from simple to complex, from technique to application, from fundamentals to their synthesis. On the contrary, from the very beginning students must face relatively complex problems in design. In addition, an ordered and structured course outline, a transparent and analytical training regimen, and clear boundaries and practical goals are the characteristics of this training method.

The Fundamentals in Visual Design course is composed of many subjects, each subject containing various steps with a progression of exercises, executed in a pattern of "lecture-exercise-critique-exercise". The goal of course lecture is to introduce the basic elements of a problem and the related content at the beginning of a new assignment, just providing which enough information for students to begin their work - it is not to indoctrinate them with content that they are not ready to accept. On the basis of their own

experiences, students will gain a deeper understanding of the content through critique of their work. Later homework will strengthen students' understanding of the content while allowing them the chance to create on their own and test their grasp of the concepts studied in class.

Educational Goals

(1) Cultivate methods of observation

During the course of study, students will learn how to distinguish the main elements of different art forms and cultivate sensitivity in their inquiries into objects of study, improving their overall ability to grasp the elements of a whole. The most important skill to be attained is the ability to discover exceptional visual phenomena

教学目的

（1）培养学生的观察方法

通过课程的学习，学生将学习如何辨别各种形式的要素，培养观察事物的敏感性，发展把握形式要素的整体能力。最重要的是掌握一种能从普通和平常事物当中发现特殊的视觉现象的能力。

（2）培养学生的绘画技巧

绘画是在纸面上的作业，教学生绘画的目的之一是给予同学对所使用的工具的信心，要求学生对不同的材料进行各种不同的实验以获得对各种可能的表达方式的了解。通过不断的学习，熟练掌握材料的运用和手眼之间的配合，通过实际解决绘画的任务而掌握绘画的策略与方法。

（3）培养学生的视觉语言

课程设置并没有为学生准备一套固定的现成的视觉语言。他们将通过对一些基本的形式要素，视觉表象和绘画策略的研究而达到对视觉问题的理解。练习的最终目的是能够在建筑设计课程中自觉运用这些视觉语言。

视觉设计基础课程与设计教学紧密结合

学习视觉设计基础课程的目的是为现代建筑等设计提供一个与之相应的感知基础，通过这一课程的学习可以丰富学生的视觉经验，增进学生的视觉敏锐性，以及提高学生运用视觉语言进行记录，表达和思考的能力。这种能力是从事设计造型的基础，有了这么一个基础，可以加强学生对现代设计理念的认识。学生在设计学习中可以更好地把握现代艺术的理念，以及抽象思维方式、空间知觉的感知、材料质感的敏锐感觉和色彩的认识分析，运用于建筑等专业设计的学习中。

视觉设计基础课程提高了学生综合设计能力

通过视觉设计基础课程的学习，为学生学习现代建筑等设计提供了一个与之相应的感知基础，学生们从自然状态形状、空间、结构、色彩等的认知、分析、研究基础上，发展出基础的设计方法与手段，在设计中对于抽象形式学习的应用与处理。空间形态的感知把握，材料质感的熟悉与运用，色彩和光影的认知与处理，提高了学生设计学习中的创造力和想象力。

through the observation of normal, common objects.

(2) Cultivate drawing techniques

Drawing will be done on paper in the form of assignments. The goal behind teaching drawing is to give students confidence with the tools that they use, and the willingness to experiment with all kinds of different materials to gain an understanding of various possible modes of expression. Through continuous study, mastery of the use of materials and hand-eye coordination will be achieved. Through solving actual drawing challenges, students will achieve a grasp of strategies and methods of drawing.

(3) Cultivate visual language

The design of this course has not prepared a specific, fixed, ready-made visual language for the students. Through the study of some fundamental formal elements, students will practice visual representations and drawing strategies and thus gain understanding of problems in visual perception. The final goal of such practice is that students will intuitively use this visual language in the process of architectural design.

The Close Connection between Fundamentals of Visual Design and Design Education

The goal of the Fundamentals of Visual Design course is to provide students with a perceptive background relevant to projects in modern architectural design. Study will deepen students' visual experiences, sharpen their visual acuteness, and improve their skills in using a visual language for making notes, expressing themselves and thinking. This set of skills is the foundation of formal design and with it students will be more able to understand the concepts of modern design. Through studying design, students will attain a better grasp of modern art, abstract thinking, the perception of space, and a keen sense of different mediums and color analysis, all of which will be of use in the study of architectural design.

Fundamentals of Visual Design Will Improve Overall Design Ability

The study of Fundamentals of Visual Design will provide students with a sensory foundation for studying modern architecture and design. Students will develop fundamental techniques and methods through the research and analysis of natural forms, space, structures and colors. In the process, they will also learn to use and manage the study of abstract forms. Skills like grasping the structure of space, the texture and use of various media, the use of color and shadow, will all improve students' creativity and imagination as they continue in the study of design.

中法联合教学的启示

赵军 东南大学建筑学院环境艺术设计系教授

东南大学建筑学院在20世纪80年代中期就提出了建筑学专业美术基础教学改革的构想，对教学计划进行了相应的调整，并在教学中进行了实践性的探索，曾经和香港中文大学顾大庆教授举办了短期的改革联合教学。从改革方案构想的提出，到不断的调整与完善，历时约25年的实践过程，其间效果如何，从已毕业学生反馈的信息来看，褒贬不一，改革后的教学方法到底对学生专业的学习与以后事业的发展产生多大影响，暂时还无法考量。

为什么要进行教学改革？教学改革的目的是什么？如何改革教学？如果不搞清楚这些问题，教学改革就可能成为一种形式，一个口号而已。

建筑是一个集历史、文化、艺术、功能、技术为一身的复杂的综合体。因此，建筑设计专业不同于其他专业，有自己的专业特点。作为一名合格的建筑师不仅需要熟练掌握相关的专业知识，更需要具有广博的知识面。

我国高等院校建筑学专业在创办初期，基本采用了以法国为代表的西方传统建筑学专业教学模式，这种传统延续至20世纪90年代。而建筑学专业的美术课教学，也基本采用了苏联（俄罗斯）列宾美术学院和巴黎美术学院的教学方法，这是由当时任课教师的知识结构所决定的。他们视建筑为艺术，专注比例、透视、空间、明暗、色彩、光和影的表现；此种以写实训练为主的教学方法完全结合了当时建筑学专业对学生培养的要求，传统的建筑设计教育与艺术训练模式对我国早期建筑设计人才的培养发挥了重要的作用，在今天仍然有借鉴的意义。

现代艺术的变革促进了建筑设计的发展，现代建筑思潮的兴起，科学技术的进步推动了建筑设计的振兴。同时，工业化、国际化又给21世纪带来了诸多新的问题，面对这些新的问题如何解决？在这种形势下，建筑学专业的教学改革成为必然，而当代建筑发展趋势中建筑学专业的美术基础教学如何适应建筑设计专业教学改革的需要，成为我们思考的问题。

美术教师作为美术基础教学改革的探索者，首先应该认识到新艺术思想的出现，成为抽象艺术和现代建筑的美学基础。而包豪斯提出"艺术与技术，一个全新的结合"的观念冲击着传统的美学思想。当今科学与技术的发展趋势，信息技术的作用不仅产生了多种风格的现代建筑，而且还产生了无形建筑，其复杂性、抽象性、非线性、以及其他科学的参与，使现代建筑表现出区别于传

Lessons Learned from Our Sino-French Joint Teaching

Zhao Jun (Professor, Environmental Arts Design Department School of Architecture, SEU)

More than a decade before the turn of the last millennium, the School of Architecture at Southeast University brought forward the idea of reforming foundational arts education for architecture majors. The teaching program was adjusted accordingly and practical experiments were conducted, as we invited Professor Gu Daqing from the Chinese University of Hong Kong to conduct an innovative short-term joint teaching project. There have been 25 years of continuous adjustments and improvements since the idea of reform was brought forward. How were the results? Looking at the feedback from graduates, they were all over the board. Whether or not the reformed teaching program had had a substantial influence on student's subsequent studies and careers, at the moment, was uncertain.

Why should we conduct reform of the educational program? What was the goal behind such reform? How should such reform be realized? If we were not clear about the answers to these questions, the program of educational reform risks would become a mere formality, nothing more than an empty slogan.

Architecture was a complex synthesis of history, culture, art, functionality, and technology. Consequently, architectural design was unique as compared with other professions. In addition to being on top of the relevant knowledge of their field, architects must also possess a wide-ranging, multidisciplinary understanding. In the formative years of the architecture major in China's higher education institutions, the western traditional architectural education made France as the main representative, was adopted which continued into the 1990s. Meanwhile, fundamental arts instruction for architecture majors largely adopted the educational methods of the Soviet Union's (Russia's) Repin Academy of Fine Arts, in addition to those of the Paris Academy of Fine Arts. This was mostly due to the structure of knowledge of the acting instructors at the time. They regarded architecture as art, focusing on proportion, perspective, space, shading, color, and shadow. Teaching methods comprised primarily of training in realistic painting and drawing epitomize the training standards of that time. In the early periods of China's architectural design talent base, traditional architectural design and fine arts training methods exercised an important influence, which continues today.

The evolution of modern art advanced the development of architectural design. The rise of a modern architectural ideology combined with the development of new technologies to impel a revitalization of architectural design. At the same time, industrialization and internationalization brought a new set of problems for the 21st century, which required a resolution. Under such a situation, reform of the educational program in the architecture major was inevitable. Within this trend of

统建筑的技术特征。近年来,有些建筑师和理论家认为应将新的科学发现,新的科学思想和世界观引入形体设计,主张以最新的科学发现为起点,积极开发建筑表现新形式,这些都对建筑设计教学提出了新的挑战。传统的建筑理论已无法适应当今建筑设计的发展要求。因此,传统建筑设计教学模式的改革对美术基础教学改革提出了新的要求。

艺术不仅给人带来美的享受,而且能陶冶人的情操,提高修养。艺术对于建筑之所以重要,并不在于造一个什么样的形体,而在于它给建筑一个什么样的灵魂。作为建筑学专业的美术教学,

我们不应该只停留在技法的教与学,而应该把它当成一种艺术认知与探索,使之成为一种激发学生建立创造精神的活动。因为当今的建筑师与设计家们更加关注纯粹美学和抽象形体构成原理及技巧;所以,如果我们按部就班地延续过去的教学经验与方法,就不可能培养出具有创造精神的未来建筑师。

东南大学建筑学院的美术基础教学改革,得到学院历届领导的重视,特别是此次中法联合教学,王建国院长、龚恺副院长在资金、教学等方面给予了大力的支持与关心。美术教学改革初期,我们在保留一部分传统教学方法的基础上,借鉴了包豪斯伊藤教授的《基础课程》教学模式,引入了平面构成、立体构成、色彩构成训练课程,强化了"结构素描"的表现。同时,结合建筑设计基础课程增加了快速表现等相关内容。与此同时,全国其他高等院校建筑学科和美术院校的艺术设计学科在造型基础教学上也进行了各具特色的改革,开设了"意象素描"、"精微素描"、

"设计素描"、"形态构成"、"形态研究"等教学改革课程。从全国高校整体情况看,美术基础教学改革还在不断的探索中。我个人认为,现阶段的教学改革还存在许多不足之处,例如:不管是写实的和抽象的造型训练,技法表现还是占了重要的部分,如何启发学生的创造性思维还没有明确的教学方法与手段,形式大于结果,造型基础训练还不能和建筑设计基础教学较好的结合等问题。

此次,我系和法国国立高等建筑学院的葛汉教授举办中法联合教学的目的,就是为了了解和学习法国建筑学校造型基础课程的教学内容与方法,为深化我国建筑设计专业造型基础的教学改革汲取可借鉴的经验。葛汉作为这次中法联合教学的主持教授,他的教学内容和方法有许多值得我们借鉴之处。例如他在讲解透视与构图时,对西方古典油画名作进行了详细的解读,他不仅从画面的主题、人物的造型、色彩与空间的表现、图底关系、光与影的表达、场景布置等等方面进行了分析,而且从画内与画

development in contemporary architecture, a core question for us was how fundamental arts education could meet the needs of the changing system of education in the architectural design major.

Instructors of fine arts who wanted to pave the way for reform of fundamental arts education should first understand the new mentality emerging in art circles, which would be able to contribute to the foundation of the aesthetics of abstract art and modern architecture. The Bauhaus' slogan, "Art and technology—a new unity" attacked the traditional aesthetic mentality. With the recent developmental trends of science and technology, the effects of information technology not only brought about different styles of modern architecture, but the creation of a new virtual architecture. Its complexity, abstraction, non-linearity, and inclusion of other sciences had separated the technological base of modern architecture from traditional architectural technology. In recent years, some architects and theorists had reflected that new discoveries in science, a new scientific ideology and world outlook which all should be applied to formal design. Their suggestions of using the newest scientific discoveries as a point of departure to stimulate the development of new forms of architecture constitute a new challenge for architectural design education. Traditional architectural theory was not equipped to satisfy the current developmental demands of architectural design. Therefore, reform of the traditional educational program of architectural design set new requirements for the reform of fundamental fine arts education.

Art not only bought people the enjoyment of beauty, but edified their spirit and enriches their awareness. The importance of art to architecture was not so much in what kinds of forms it created as what kind of spirit it brought to the discipline. Fundamental arts education for architecture majors should not be limited to the study of technique. It should continue to explore a new perception of art and become an activity that sparkled students' creative spirit. As contemporary architects and designers put more importance on pure aesthetics and the principles and techniques of abstract forms, if we routinely extended traditional educational experiences and methods, we would not be able to cultivate a new generation of architects who possess the creative spirit.

Reform of fundamental arts education had been stressed by the previous leadership of Southeast University's School of Architecture, and this instance of Chinese-French joint teaching was a prime example. President Wang Jianguo and Vice-president Gong Kai gave tremendous support for the project in aspects ranging from funding to instruction. During the beginning stages of our reform of the fine arts, while retaining some of the traditional educational methods, we borrowed from Bauhaus professor Johannes Itten's Design and Form: the basic course at Bauhaus, and included courses of training in the structure of planes, solids, and colors to strengthen the quality of our "structural sketches". In addition, the integrated architecture and

外（观赏者）的呼应关系做了场景与形态上的分析。他对西方名画并非单纯从赏析的角度进行解读，而是通过理性的、科学的研究阐述绘画的内在规律和艺术精神。现代抽象艺术也是葛汉教授讲授的重点内容，他在介绍抽象艺术的美学形式与构成原理的同时，还着重阐述了当代艺术视野下的建筑设计与创作的发展趋势。

通过中法联合教学，以及葛汉教授教学经验的介绍，法国高等学校艺术基础教学特点表现为：从传统文化中引入思考，通过艺术作品的欣赏与分析提高学生的修养，拓展学生的思维，以及艺术观的自我培养与批判精神的建立；树立科学的思想，启发学生的创造性活动；强化设计思维的训练。他们的教学内容和方法是跨时空，多维度、开放式，多学科融合、重启发、鼓励创新，这些对我国建筑学专业与艺术设计专业的美术基础教学都有着重要的启发与积极的借鉴作用。

design curriculum added quick impressions and other similar exercises. Meanwhile, institutions of higher education across the country were implementing their own brands of educational program reform in the departments of architecture and arts & design. They had opened innovative courses including, "interpretive sketching", "microscopic sketching", "design sketching", "the structure of form", and more. Looking at the situation of higher education across the country, it was clear that fundamental arts education was in a continuous state of discovery.

In my personal opinion, the current stage of educational reform still had many inadequacies. For example, no matter whether in training of realistic or abstract forms, technique was still given the strongest focus. Clear methods to incite students' creative thinking still had not been established. With means eclipsing the end, foundational training in forms could not achieve a satisfactory integration with education of the foundations of architectural design.

My department's recent joint teaching cooperation with Professor Guerin, from the Architectural National Superior Schools of France, had as its goal to study and understand the method and content of the curriculum of foundational forms education in French architecture schools. The experience gained could then be used to deepen China's reform of foundational education in architecture and design. As the leading professor of this joint teaching project, Professor Guerin provided much in his methods and content that we could learn from. For example, when explaining perspective and composition, he utilized many classics of western painting to aid his lecture. Not only did he analyze aspects like the topic, subject, use of space and color, the relationship between figure and ground, the expression of light and shadow, and layout of the works, he also carried on analysis of the scenes from the perspective of the interactive relationship between the works and their audience. Moreover, he did not carry out his interpretation of these works solely from the perspective of appreciating their appearance. Rather, he used scientific research to explain the principles and artistic spirit behind them. Modern abstract art also had particular focus in Guerin's lectures. At the same time as introducing the aesthetics and compositional principles of abstract art, he described contemporary art's vision of architectural design and its developmental trends.

From this joint teaching experience, and based on the introduction given by professor Guerin, teaching of the fundamentals of fine arts in France's higher education institutions could be characterized by the following points: beginning inquiry from the perspective of traditional culture, using viewing and analysis of works to improve students' awareness, developing students' thinking skills, and cultivating a personal conception of art while establishing habits of critical thinking; establishing scientific thought, inspiring creative activity; training to strengthen design thinking. Transcending space and time, their educational content and methods were multifaceted, open, multidisciplinary, inspirational, and encourage creation. These qualities would prove to be a source of inspiration and an important reference as China continued to reform its foundational arts education in the curriculums of architecture and arts & design.

超透视表达
——对建筑绘图和设计思维的有益补充

沈颖

建筑师用图纸的方式来思考、表达、呈现和解读设计,作为设计意图的载体,绘图具有权威性和简明性的特点,它拒绝产生歧义和由于误读而造成的含混不清。绘图是实现空间描绘的主体,在专业教育中,也有相当的内容是关于如何运用这种主流的媒介去表达和交流,并在此基础上锻炼专业化的对空间的思考方法。自文艺复兴始,建筑师的专业地位通过绘图而得以确立,绘图成为完成脑力和体力社会分工的度量。然而,绘图同时是一个独立观察和思考世界的方式。它的系统化、系统的成熟化与普遍化以及对于视觉空间的遵循是否在某种程度上也会束缚对空间的思考,是否在平面、立面、剖面的逻辑之侧有其他元素有益于空间的设计及其过程的表达?

(1) 建筑绘图的历史脉络

古埃及的画师将他们的所见如实地再现。如果是方形的池塘,就会以方形还原在画面上,如果树木是等距离的,那么在画面上无疑也是以等距的形式出现的,人物也以最易识别、最直观的侧影形式被描绘出来。当时的绘图与其说是一种对视觉空间的模仿还不如说是一种对直觉的还原。在中世纪的欧洲,建筑的建造往往是数代人的努力积累而成的,其间,业主、设计方和施工方也历经数代,建筑不是某个人或某个团队的设计。工匠们直接用建造和模型来沟通意图,而不是事先绘制设计。作为推敲建筑细部的图纸绘制在羊皮纸上,用完的图纸被打磨掉以供再次使用,而不是用以保留建筑设计方案。单一的设计概念和现代意义上的建筑师都还不存在,成系统的建筑绘图方法也未诞生。

到了文艺复兴时期,"设计图"的概念被提出来,它成为采用绘图形式、区别于绘画、表达设计的特殊范畴。设计师把将要建造的建筑的特征和信息通过绘图抽象地转述出来,这已经与中世纪工匠所理解的绘图不一样了,设计图也成就了设计师这一行业,建筑绘图成为设计实践的基

荷兰建筑师维尼·玛斯以多媒体装置的方式,建立了一个虚拟的Datatown,将数据信息转化为空间形体,探讨密度问题的同时呈现出数字化的美学

A Perspective Expression: An Instructive Complement to Architectural Representation and Design Thinking

Shen Ying

Architects use blueprints to ponder, express, represent, and explain designs. Architectural representations are unique in their authority and simplicity, refusing to engender multiple interpretations, and thus eliminating the confusion brought on by a misreading. Drawing is the main vehicle for portraying space, and professional education contains a large body of content regarding how to use this mainstream medium to enhance expression and communication, building a foundation for exploring specialized conceptualizations of space. Ever since the Renaissance, the professional status of architects was established through their sketches, and sketches became an important measurement factor in the division of labor between intellectual and physical work. However, drawing also stands alone as a unique method for thinking about and exploring the world. Can the system of drawing with its allegiance to the visual perception of space, as it matures and is inseminated, limit our thinking about space? Can thinking about space, whether on a plane, in three dimensions, or in cross-section, be improved through considering other elements relevant to the design of space and the methods of expression used in the design process?

1. The History of Architectural Drawing

Visual artists of ancient Egypt sought to accurately reproduce the things they saw. If it was a rectangular-shaped pool, they would use a rectangle to reconstitute it on their tableau. If trees were equidistant from each other, there was no doubt that they would appear equidistant in their representation. People were portrayed in profile, the easiest method to distinguish and recognize. Drawing at that time was not so much imitation of visual space as a trigger of intuition. In Europe during the Middle Ages, the construction of architecture was made through the accumulation of many generations of labor, as customers, designers and builders often spanned many generations. Architecture was not according to the design of a single person or team. Craftsmen used direct construction and models to communicate their intent, rather than drawing their designs beforehand. Especially elaborate parts of the buildings were drawn on lambskin, but when the drawing was not needed it was polished away so the skin could be reused, so there was no record to preserve architectural blueprints. A single concept of design, the architect in its modern sense, and a systematized method of architectural representation still did not exist.

The concept of the design drawing came out during the Renaissance. It defined the scope of architectural work with its method of mapping, distinguished from those of drawing or painting. Designers used abstractions to relate the important information and features of the building to be constructed. This was also a distinct divergence from the concept of design drawing in the Middle Ages as

础。1563年,乔治·瓦萨里(George Vasari)创立了佛罗伦萨艺术学院,他主张将雕塑和建筑都归为设计艺术。在那里画家、雕塑家和建筑师们通过绘图来得到训练。这种教学方式将设计作为一种脑力劳动从建造的体力劳动中分离出来。经过文艺复兴的变革,建筑师从直接的建造抽离出来,投身于概念化的图形表达。

之后的巴黎美术学院采用设计几何学的教学体系,强调对于建筑几何属性的认识,平面、立面和剖面是分析建筑几何属性的有效途径,它成为建筑教学体系的权威。直至20世纪初,这种教学思路才被包豪斯的教师们试图予以突破和更新,将抽象的绘图与实在的材料、将设计与工艺结合起来,鼓励在工作室中体验建造与材料,反对建筑的美术化倾向。而设计几何学体系的权威地位则不会因其存在的些许不足而失效。

(2) 透视思维的演进

透视科学提供了一个让人把眼前显现的形状画出来的手段和方法,它表达了某个形状在空间中的相对位置所导致的视觉上的变形。透视画法(perspective representation)在艺术界、设计界取得合法的地位是一个里程碑。它的诞生是人类文明史上一次重要的空间意识的变革,它对意识形态起到深远的影响,这个影响一直延续到当今的建筑及其他艺术与设计实践中。它鼓励了对三维深度空间和单一空间的稳定的表达。同时,空间和时间被分化成两个独立的概念,空间意识凸显,时间成为附属于空间的概念。20世纪,自立体主义始,新的媒介和空间意识开始萌芽,透视作为绘画和设计表达的主导媒体,开始受到些许质疑。阿恩海姆(Arnheim)将古埃及的绘画和文艺复兴的线性透视作了比较,将图和空间意识之间的紧密关系揭示出来。他认为埃及人的绘图法和线性透视由于对空间的认识不同,属于两种截然不同的再现空间的方式。前者是对事物"本来面目"的再现,后者是对视觉呈现出来的状态的再现。古埃及人如实表达事物的属性,而不像线性透视那样对视觉空间进行模仿。20世纪存在主义雕塑家贾科梅蒂(Giacometti)以"瞬间叠加"的雕塑方式去"抓住并揭露一种真实",他把一个个泥稿似的人物改来改去,旨在捕捉和表现人的一种真实存在。"所以贾科梅蒂的'苦'首先不在技法,不在做,而在'看'",就是要捕捉到一种印象而不是一般意义上的观看。他所表现的真实是由无数瞬间印象的显现而叠加在一起得来的。他的雕塑从而存留着印象的"看"的过程,而不只是印象的结果。

古人和今人的视觉感受并无二致,但在认识上却不能说没有区别。贡布里希(Gombrich)在《艺术与

构成主义艺术家罗德琴科对不同视点的捕捉所作的蒙太奇实践,画面具有多重释意

designs became the essence of the design industry, and architectural drawing was its foundation. In 1563, George Vasari founded the Florence Academy of the Arts of Drawing with the vision of classifying both sculpting and architecture as design arts. Accordingly, the academy used drawing to train painters, sculptors, and architects alike. This method of education served to separate intellectual work from manual labor, with the Renaissance welcoming a new kind of designer completely separated from the construction process, who only needed to focus on conceptualized graphic expression. Afterward, the École Nationale Supérieure des Beaux-Arts in Paris used an educational system based on geometric design, emphasizing the necessity of geometric knowledge in architecture. In this system, planes, elevations, and cross-sections were the effective channels for analyzing architecture, which eventually became the standard in architectural education. This lasted all the way until the beginning of the 20th century, when the Bauhuas school tried to reform it by combining abstract drawing with the medium, encouraging experiments with materials and construction in the studio, thus combining the work of designer and craftsman and moving away from the trend of fine-arts in architecture. However, in the face of these apparent inadequacies, the authority of the geometric design system was not lost.

2. The Evolution of Perspective

The use of perspective allowed artists to draw the objects before their eyes with such a technique that they would convey the relative location of the objects in space by capturing the corresponding changes that took place in one's visual perception. Perspective representation's status as a law was a milestone in art and design circles, and its emergence represented an important revolution in human civilization's thinking about space. Its ideological influence had been vast, extending into contemporary architecture as well as other areas of arts and design. Perspective representation encouraged a stabile representation of three dimensions in one-dimensional space. At the same time, space and time became two separate concepts, with space being highlighted and time existing as a secondary concept. In the 20th century, with the advent of Cubism, a new attitude towards space and medium began to sprout, and perspective as the methodological foundation of drawing and expression of design was called into question. Arnheim conducted a comparison of the lines of perspective between ancient Egypt and Renaissance art and revealed the intimate connection between art and the concept of space. According to his research, the ancient Egyptian method of drawing and the linear perspective method were two entirely different methods for representing space, and represented two distinct ways of thinking about it. The former focused on representing the original features of the objects, while the latter focused on representing the state produced by visual perception. Ancient Egyptians accurately represented the characteristics of objects without using the technique of perspective as a tool for imitating space. 20th century existentialist sculptor Giacometti used a spontaneous method of sculpting to capture and expose a certain truth. By contorting clay people into various positions, he searched for a way to express truths of people's existence. So, Giacometti's angst was not in technique. It wasn't in doing, but rather in seeing, in the sense of pursuing a certain impression or feeling. The truth he produced was obtained by superimposing countless spontaneously-created impressions. Thus, his sculpture left behind an impression of the process of seeing, rather than the "seen" results.

错觉》中将知觉的经验构造主义理论运用到绘画的再现问题中来。他围绕这个理论对在西方传统上关于"视觉的"和"认识的"两个概念之间的区分展开论述。他认为"视觉"等同于对感觉和视网膜影像的意识;"认识"负责处理关于知觉和客体的假设的问题,只有用这一种方式才能给予感觉上的混乱以秩序。波兰思想家让·盖伯瑟(Jean Gebser)将人类的意识发展历程总结为三个阶段:透视之前(pre-perspective)、透视(perspective)和超透视(Aperspective)。在透视诞生之前,一种总体的、直觉的思考方式占主导地位,自然独立的空间意识还尚未形成。透视将空间和时间分离,时间维度退隐,形成单一静止的空间感,视觉感知成为空间体验的主导方式。盖伯瑟还提出了超透视的新可能,它并不是走向透视的反面,而是一种新的态度。时间成为第四维,来补充原来主导的空间意识,由此形成时空意识的重新结合,更新原有的静态的空间意识。每个单独的观察者在不同的时间也会有变化的观察,同一个空间中纳入了不同的视点,

也渗透入不同的解释,这些视点与解释重组,从而形成多重阐释。这三个阶段在时间上不是新老交替而是同时并存的关系,但他们具有鲜明的时代性,在某种程度上是共存的,同时又带有各自所主导的那个时期意识形态的印记。

(3)超透视表达的可能

从胡塞尔(Husserl)现象学的观点看,以实证主义为基础的现代科学排除了它自身现实发生的境域,改变了传统科学对境域的依存,使自身成为绝对客观性的与直观境域无关的知识体系。基于这个知识体系的建筑思维,同样缺乏与人文的融贯性。这种线性思维缺少整体性,在追求效率与效益的同时容易疏忽各种活跃元素与能量组成的自然的多样性,视自然为承载建筑的背景,将人视为生活在其中的抽象元素,忽略人类行为的多样性。

盖伯瑟并没有将透视作为一种绘画方法,而更多地将其看作是一种认识和思考世界的方式,一种带有极强时代特征的空间意识和体验方式。在

透视之前,运动和行进是体验空间的重要方式;之后,透视的视觉范式取得了霸权地位,生理与心理空间被转化成一种数学空间,空间与体验相脱离;当透视空间进一步发展,则诞生了一种更具现代性的空间意识:一种体现空间位移、主客体间交互的时空观。

在我们选择利用透视来再现空间时,在我们感叹透视这种方法带来逼

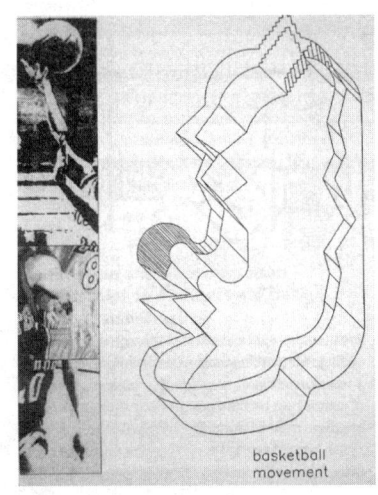

利用一个篮球的定点位置和移动轨迹概念化完成一个图形结构

The visual perception of ancient people and that of people today were no different from one another, but on the level of cognition it could not be said that there is no difference. In ***Art and Illusion***, Gombrich applied a theory of the construction of conscious experience to drawing and the problem of representation. He focused his theory on traditional western ideas of the "visual" and the "cognitive" and discussed their differences. He believed visual perception consistd of the feeling and consciousness of a retinal image. Consciousness was responsible for managing the gap between objects and visual perception, for only through such a means could the perceived chaos be given order. Polish thinker Jean Gebser summarized the evolution of human consciousness in three stages: pre-perspective, perspective, and Aperspective. Before the birth of the perspective world, an all-encompassing, intuitive way of thinking was dominant, and natural independent space still had not come into being. Perspective brought a division between time and space, and the pushing away of the time dimension brought about a single, static sense of space, and visual perception became the main object of spatial experience. Gebser also brought up the new possibility of Aperception, not so much a negation of perception as a brand new perspective. Time became the fourth dimension in order to bring balance to the dominant spatial perception, thereby bringing about a recombination of space-time consciousness, revising the previous perception of static space. Every individual observer at different points in time would have different perceptions, uniting different points of sight onto the same space, saturating it with different possible explanations. The different combinations of viewpoints and explanations made for multiple interpretations. These three modes of consciousness were not developing in time from old to new, but simultaneously coexist. However, they could reflect the distinctiveness of the times. To the degree that a time period expressed each mode of consciousness, the time would be imbued with the stamp of its ideology.

3. The possibility of Aperspective Representation

From the standpoint of Husserl's phenomenology, modern science based on positivism altered traditional science's over-reliance on its realm by delineating its own space of existence. It created a system of knowledge based absolutely on objective and perceivable phenomena, regardless of the realm in which they occur. Architectural thinking based on this system of knowledge naturally lacked coherence with the humanities, for a linear structure of thought los the capacity for holistic understanding. While pursuing efficiency and achievements it was easy to ignore the natural multiplicity comprised of dynamic elements and potentials, to look at nature as merely the background for architecture, to look at people as abstract variables that merely inhabited them, to ignore the diversity of human behavior.

Gebser never described perspective as a method for drawing, and rather regarded it as a method for thinking about and understanding the world, a consciousness and experience of space, with strong characteristics of different time periods. In the pre-perspective world, motion and progress were the main modes of experiencing space. Afterward, in the perspective world, visual perception reigned supreme, transforming the space of biology and psychology into one mathematical space. Space and experience were separated. With the further development of perspective space, an even more modern consciousness of space would develop, in which space shifts, and

真的再现的空间深度的同时，它已经限定了建构空间和体验的逻辑。这种逻辑带有明显的时代特征，它并不是一成不变的真理，随着空间意识和体验在历史发展中的不断进化，绘图也会不断地进化。在一系列专业化的过程中，传统的建筑绘图不足以触发新的设计思维；对绘图的工具性的强调和训练，忽略的是它作为体验手段的作用。绘图不仅仅是空间意识的被动表现，同时也可能是新空间意识诞生的催化剂。超透视的表达可以将绘图和空间体验重新连接起来，使绘图参与到认识、解读、构思、生发设计的过程中来。

由此看来，仅仅沿承经典的空间构成章法和传统的审美法则，对于空间的创造来说是不够的，需要通过身心的体验获取空间塑造的灵感，敏感地捕捉周边存在的特殊性。从更具现代性的空间意识出发，当生活空间被给予与抽象空间等同的重视时，建筑的地域性和特殊性等特质也会自动生成。建筑师如其他领域的设计师一样，应力求体现人类生活的本真状态，而生活的本真性是一个充满变化的、动态的生活过程。赖特（Sharon）对业主的委托以充分考虑业主的生活方式为原则；夏隆注重事件的本质，认为人类与变化的时空相连，自然是非静态的，坚持建筑创造必须具有文化价值，坚持由内而外的设计，以自然结构形式去对抗理性的几何思考。NL建筑师事务所的Blok K住宅通过对空间别样的分割带给人们新的居住方式，将中廊对角线状布置，所有居住单元的长度和高度都经过不同程度的拉伸与压缩，仍保持每套单元足够的体积，住宅楼的标准模式因此而改变，这种具有弹性的分割创造了一个独特的居住建筑。

Clare Robinson在*Browsing, Bouncing, Murdering, and Mooring - Negotiating the Relationship between Inhabitation and Representation*一文中列举了多个建筑绘图的实例，探讨了关于居住空间、生存环境及想象空间之间的关联。文中的研究和实验延伸了空间与事件的关系，传达了空间形态存在的意义，如：身体与图书馆的书架与书本，篮球与运球者，谋杀者与居住空间，停泊的船只与旧金山海岸线，两两的相互关系也呼应了真实空间与想象空间的关系。概念性建筑可以作为居住与想象之间的再现形式，随着环境或历史事件的介入，建筑的空间结构原本的功能与活动会因此而被取代和转化，而居住与设计的关系也融入各种可能的空间、事件、移动中。随着空间里被关注的对象不同，居住模式和生活形态也随之多样化。

建筑绘图不仅可以是对现实建筑的描绘，也是对未来建筑建成以后情况的一种预测，建筑师通过绘图描绘概念，将概念视觉化。作为一种比较专门的图像，建筑绘图也是一种具有社会性的符号化了的现实。作为概念和建成空间之间的媒介，它主动地对其所传达的建筑概念进行过滤，采用不同的具体方式以使一些信息得以识别和传达，而另一些信息得以阻断和删除，借此激发新的建筑空间概念。因此，对超透视表达的探讨有益于补充和完善建筑绘图的工具性与有效性，也有益于设计思维的生发。

subjectivity and objectivity interact in the perception of space-time.

When we chose to use perspective to represent space, as we gasped at the true-to-life representation of space that it created, it had already qualified the logic of experiencing and constructing space. This logic could define the characteristics of a certain era. It was not an unvarying truth, but rather changed with the consciousness of space and experience in the flow of history. Thus, the concept of drawing also changed throughout history. In a series of professional movements, traditional architectural representation was not enough to sparkle a new mode of thought in design. During adherence to certain tools and educational practices, what was lost was its value as a method for experience. Drawing was not just the passive manifestation of the consciousness of space. It was also the catalyst for the construction of a new mode of spatial consciousness. Aperspective representation could bring drawing and the experience of space together once again, allowing drawing to be a determining participant in understanding, interpreting, conceiving and developing the process of design.

From this we could see, merely extending the classical composition of space and aesthetic principles was not enough to create a new space. Creation needed inspiration gained from human experience, and the sensitivity to seize the uniqueness in one's surroundings. From the perspective of a more modern consciousness of space, when the space of life was given the same recognition as abstract space, the regional and unique aspects of architecture will be born naturally. Just as other designers, architects must do their best to embody the essence of life's condition in their work, and the condition of life was an ever-changing, dynamic process. Wright took constant reference to his clients and ample consideration of their lifestyle as his principles. Sharon emphasized the importance of the quality of the event, believing that humanity and the changing times were linked, that nature was not static, and that a lasting architectural construction must have cultural value. With this he used a natural structure to oppose rational geometric thought. The Blok K dwelling designed by NL Architects used different divisions of space to give people a new style of living. With design based on a central diagonal dividing line, each living unit's length and height underwent a different degree of expansion or reduction, all the while preserving adequate size of each unit. Consequently, the standard unit was indefinable. This kind of elastic division was a unique creation of residential architecture.

In the article, "Browsing, Bouncing, Murdering, and Mooring" — Negotiating the Relationship between Inhabitation and Representation", Clare Robinson listed many examples while discussing the connection between living space, living environment and imaginary space. The research and experiments in the article extended the relationship between space and events while conveying the significance behind the configuration of space. For example: the body and a bookshelf in the library, a basketball and the person dribbling it, a murderer and the living space, an anchored ship and the coastline of San Francisco; pair by pair of correlated items echo the relationship between real and imagined space. Conceptual architecture could be a method for representing the space between inhabitancy and imagination. With the involvement of the environment or historical events, the structure of architectural space and its functionality will undergo corresponding replacement

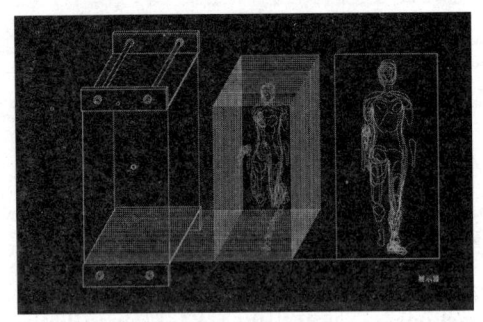

余海男同学的习作通过运动来揭示时间向度的存在,以此诠释"建筑是对运动的框景",叙事的体验由时空二者共同构成

(4) 2010年中法联合教学对于超透视表达的探索

2010年度的这次中法联合教学为即将升入大三的学生在如何观看如何思考方面提供了很好的学习机会,葛汉老师对西方艺术特别是对艺术作品中的空间意识进行了介绍,以具有过程性教学特点的"空间中的一个点"、"没有什么比一个表面更有深度"、"所有建筑都是对运动的一个框景"等先后三个主题对"观看"进行了梳理,以此作为学生思考的出发点,试图引导学生从一个起初与实体空间并非具有直接关联的想法完成向作品的转化,并鼓励学生将实践成果与未来的专业设计课题相结合。学生也由此有机会从透视绘画实践中片刻抽离出来,从各自不同的角度深入思考,找出各自与此相契合的点,并培养起对作品的独立思考的能力和批判性精神。

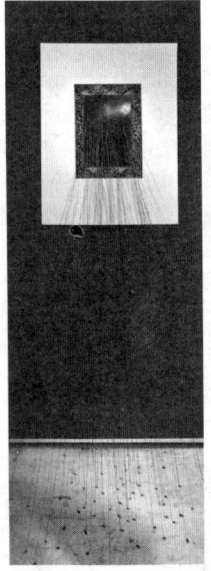

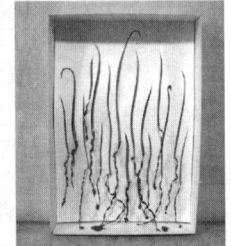

唐晓兰同学将撕裂动作理解为一种对平面的突破,在平面内与外的关系上进行探索,围绕三个主题的系列作品体现出前后的延续性

参考文献

[1][美]阿恩海姆.艺术与视知觉[M]. 滕守尧,译. 成都:四川人民出版社,1998.
[2]黄琪,[瑞士]贾科梅蒂[M]. 北京:中国人民大学出版社,2004.
[3][英]贡布里希.艺术与错觉[M]. 林夕,等译. 长沙:湖南科学技术出版社,2009.
[4]Jean Gebser,Noel Barstad, Algis Mickunas. The Ever-Present Origin[M]. Athens:Ohio University Press, 1984.
[5] Clare Robinson. " Browsing, Bouncing, Murdering, and Mooring"—Negotiating the Relationship between Inhabitation and Representation[J]. Journal of Architectural Education, 2005.9(1)

and transformation, and the relationship between inhabitancy and design would enter different possibilities of space, events and movement. As the focus within space changes, living patterns and modes of life would multiply.

Architectural representation could be a portrayal of present architecture as well as a prediction of the future of new architecture to be built. Architects used representations to depict a concept by making the concept visual. Such specialized images were also symbols of society. As a medium between the concept and the completed space, the mode of representation filtered the concepts that it could communicate. By using various methods to allow certain information to be distinguished and communicated, other pieces of information were blocked or erased, thereby inciting the creation of a new concept with which to view architectural space. Consequently, the exploration of Aperspective representation was beneficial for supplementing and perfecting the tools and efficacy of architectural representation and the development of design thinking.

4. The 2010 Joint Education Experiment into Aperspective Representation

2010's joint-education with France was a great opportunity for our rising third-year students to study how to observe and think. Professor Geurin introduced western art, especially the concept of space in various works. In a series of 3 lessons, "one point in space," "there is nothing deeper than a plane," and "all buildings are windows to movement," observation was made the main object, and students were provided with a starting point for reflection. The goal was to lead them to start from an idea that did not necessarily have a direct connection to substantial space and to transform it into a work, while encouraging them to integrate their projects with practical applications and future problems of professional design. Students thus had the opportunity to make a brief departure from perspective representation, and from their individual perspectives reflect and find a point of accord within themselves. In this process they cultivated their skills of independent interpretation and their critical spirit.

References
[1] Arnheim, Rudolph. Art and Visual Perception[M]. translated by Teng Shouyao. Chengdu: Sichuan People's Press, 1998.
[2] Huang Qi, Giacometti [M]. Beijing: The Press of People's University of China, 2004.
[3] Gombrich, E.H. Art and Illusion[M]. translated by Lin Xi et al; Changsha: Hunan Science and Technology Press, 2009.
[4] Jean Gebser, Noel Barstad, Algis Mickunas. The Ever-Present Origin[M]. Athens: Ohio University Press,1984.
[5] Clare Robinson."Browsing, Bouncing, Murdering, and Mooring — Negotiating the Relationship between Inhabitation and Representation," [J]. Journal of Architectural Education, 2005 (1).

心智与图式

朱丹

设计往往被看成是一个灵感突现的过程，然而事实上，这种看似偶然的"灵光一闪"往往来自于设计师日常经验积累。任何设计思想其实都是他们脑海中已经存在的原始材料的另一种表达方式，这样便形成了对事物的全新认识。通常，积累越多，认识越广，可能出现灵感的机会就越多。积累起来的经验影响到人们的思维，最终形成个人具体的心理特点与心理规律，我们将之称为心智。一个心智成熟的设计师必然拥有极为丰富的经验或者那些他们可以直接或间接从其他地方获得的与其专业相关联的东西，将之联合碰撞才能获得创新上更广泛的自由度。任何在平淡的生活中的所见所闻，以及从别人处获得的好的想法对于设计师而言都是相当重要的。他必须注意并且思考从而发现意义，将这种思考的结果以图形和具体的设计样式（我们称之为图式）表现出来，这就是设计师与常人的区别之处。

在谈论人们认识与体验世界的方式时，我们就会涉及"感知"这个概念。"感知"实质上包含了感觉与知觉两种不同的心理现象。"感觉"是人脑对于事物个别特性的反应。纯粹的感觉可以被理解成为一种未分化的，转瞬即逝的点状冲击感受，是局部而片面的，故感觉与真正的体验往往并不相符。"知觉"则是人脑对于一个具有某些统一特征的对象所产生的反映，它包含着对于事物总体特征的理解，是整体的和较为深入的，它形成了人们心智的基础。但是，人们总是有意无意地选择少数事物作为知觉的对象，而其余的事物则作为模糊的背景而存在，因此，这种基本的知觉已经具有了一种"意义"，即我们只知觉和体验我们自认为有意义的事物。其中，事物意义来自于当事人过往体验的回忆与经验，这些回忆与经验便构成了一个人的心智。显而易见的是，每个人具体的生活经历、文化背景、兴趣爱好等要素各不相同，因而对相同物体产生不同的关注。简而言之，外部世界的原本（text）不是被复制而是被重新整合了，不同的心智状态和观看的心理规律支配了我们所能看到的具体内容。比如，漂亮的雏菊的花头向我们展示出两种精美的图式，一是较为容易观察到的放射状图式，花瓣从花蕊的边沿放射出去，使人们联想到太阳耀眼的光辐射线条；另一种来自于花蕊，这是由两个同一中心的曲臂围绕着中心旋转而成的。注意到第二种图式的人与第一种图式观察的心理基础并不相同，第一种图式的观察基于将整个雏菊的花朵当成图形来感知；而后者则是进一步将花

Psyche and Schema

Zhu Dan

Design is often regarded as a process that involves the sudden appearance of inspiration. However, this seemingly occasional "flash of spirit" usually comes from the accumulation of everyday experience on the part of designers. As a matter of fact, every design idea is just another mode of expression of the original material existing in their mind, thus giving rise to a brand new understanding of things. In general, the amount of accumulation and the extensiveness of understanding are in direct proportion to the probability of the emergence of inspiration. The accumulated experience affects people's thinking, and eventually gives rise to concrete psychological characteristics and individual personalities, which we call "psyche". A designer with a mature psyche must have very rich experiences or things associated with his/her profession that he/she can directly or indirectly obtain from other sources. A more extensive degree of freedom in innovation can only be attained by combining them together. Therefore, what is heard and seen in ordinary life, and all the good ideas obtained from others are very important for a designer. He/she must pay attention to and ponder them in order to discover their meaning and express the result of this thinking in the form of pictures and concrete design patterns (which, together, we call schemas). This is what makes a designer different from other people.

When discussing people's modes of understanding and experiencing the world, we will touch on the concept of "perception". In essence, "perception" includes two different mental phenomena, "sensation" and "consciousness". "Sensation" is the reaction of the brain to individual characteristics of things. Pure sensation can be understood as an undifferentiated and instantaneous feeling of impact, which is local and bilateral; therefore, sensation and actual experience do not always conform to each other. In contrast to sensation, "consciousness" is the reflection of brains to a target with some uniform characteristics and comprises the understanding of the general characteristics of things. As an integral and relatively deep entity, "consciousness" forms the foundation for people's psyche. However, people always choose a small number of things as targets of consciousness, while other things exist vaguely in the background. Therefore, this basic consciousness already has some "meaning", namely, we only perceive and experience things we deem meaningful. The meaning of things comes from a person's memories and experience from the events of his/her life. These memories and experiences constitute this person's psyche. Obviously, different persons have different concrete life experiences, cultural background, interests, hobbies and other factors, making them place different values on the same objects. In short, the text of the outside world is not copied but re-integrated. Different states of psyche and psychological laws

瓣作为背景，花蕊作为图形来感知。然而，一旦发现了这种旋转状的图形，我们便能在其他物体上轻而易举地找到相同的图式，比如飓风、大洋的潮流、植物萌芽时卷曲的叶子等等。数学家通过这种图式发现了斐波纳契数列，而艺术家们则在这个图式中抽象出黄金分割的美学样式、建筑师则将这种图式用于建筑造型之中。在这个不断深入的观察体验的过程中，人们的心智得以不断地提高；反之，成熟的心智又拓宽了我们观察世界的视野。

艺术家和设计师总能比常人发现更多有趣的图式，这并不是说他们的眼睛有异于常人，而是他们的工作长期与视觉图像打交道，这使他们对图形的敏感度高于常人，对于一般人而言的普通形式在设计师眼中可以重新转化为另一种意义的图式，它取决于在某种特殊情况下通过由形式引起进而与形式产生联系所表达的意义。在"时光的表面"这一作业中，学生在偶然的情况下发现了长期受热及摩擦形成的餐桌表面有趣的痕迹。这有可能是他们在无数次观察到这一肌理中的极为平常的一次，但在"没有什么比表面更有深度"的课题引导下，观察者的态度发生了改变，图形从它们原有意义链中被释放出来，从而失去了本来的意义，同时在一个新的背景下被观察，唤起了其它意念，并形成了别的什么东西，可以说形式自由了，它从更早的含意体系中被分离，自由地担任了新的角色。

然而日益增长的经验也会限制我们的开放性思维，成为创新路上的绊脚石。创造力就意味着设计师的观念与概念与众不同，这来自于对旧价值观的批判，你必须尽可能地开放你的思维，并惯于怀疑旧的概念或"真理"。然而做到这一点并不容易，在我们处理一项任务时，旧的观点和思想会作为一种经验自然地涌现出来，你必须时时提醒自己摆脱它们并不断尝试反问。课题"没有什么比一个表面更有深度"的关键点是对于"表面"概念的理解，多数人的惯性思维围绕着物体外皮所呈现出来的图式展开设计。然而，当你从一个截然相反的角度来看待表面，将表面理解成为事物的发展过程的某一个片段时，就会发现隐藏于表面之下的东西更可以反映事物的深度和过程性。季欣和张翔的作品中就表现出设计者强烈的反思"表面即表皮吗？""表面是真实的吗？"第一个作品将剖面作为表面来理解，用此来反映事物的内部结构和肌理，这样的肌理和被剖的物体有着密切的联系，通过剖面的表达可以反映事物的本质。而第二个作品"具有欺骗性的表面"进一步揭示了表面的迷惑性，有时表面可以被理解成装饰，无法反映事物本质，而内部的结构是有待揭示的。作品希望展现揭示过程的震撼力，因此设计思路便产生了，通过一种"野蛮"的方式表达的手法强调了表面与本质的对比。这里，作品所展示的并非一个简单的结果，而是透过图的作者对这一问题的与众不同的视角与思考的过程。知识总是使人们趋向于寻找可认知的部分，因为这部分是可以被人们尽快地解读和理解的，但若只停留在这一步，心智便不会得到进一步的机会。你越是对固有意义表示怀疑，就越容易对此进行分析，这样思维便活跃而

of observation govern the details we can see. For example, the beautiful blossom of a daisy unfolds two kinds of elegant patterns: the first is the radial schema that can be discerned easily; the petals radiate from the edge of the pistil, reminding people of the dazzling optical-radiation rays of sun; the other comes from the pistil: two homocentric arms revolve around the center. Those who notice the second schema have a different psychological foundation in comparison with those observing the first. The observation of the first schema is based on the fact that the entire flower head of the daisy is perceived as a picture; in contrast, the latter perceives the pistil as a picture with the petals as the background. However, once this revolving picture is discovered, we can find the same schema easily on other objects, e.g. a hurricane, the ocean tide, the curly leaves of a plant in bud. A mathematician discovered the Fibonacci series through this schema, an artist abstracted the aesthetic pattern of the golden section in this schema, and an architect applies this schema in design. In this continuously deepening observation and experience, people's psyche is continuously enhanced. In turn, a mature psyche can expand our perception of the world.

Artists and designers can often find more interesting schemas than ordinary people, which does not mean that their eyes are different from those of ordinary people, but that they frequently deal with visual images during their work, so they have a higher sensitivity to pictures than ordinary people. In the eye of designers, forms that seem ordinary to common people can be re-transformed into new meaning in a new schema, which happens in those special circumstances when a perceived form induces a connection in the psyche and meaning is expressed through that connection. In the exercise " the surface of time" during the recent joint teaching camp, students happened to discover interesting traces on the surface of dining tables arising from long-term heating and friction. This is possibly just one among countless occasions when they noticed this texture. However, guided by the exercise "nothing is deeper than the surface," the attitude of the observer undergoes a change and pictures are released from their original chain of meanings at the same time, observed against a new background, they evoke other thoughts and give rise to new impressions. In other words, the form becomes freer, and it freely plays a new role after being separated from an earlier system of meaning.

However, ever-growing experience can also restrict the openness of our thoughts and become an obstacle on the road to innovation. Creativity means that designers have unique concepts and ideas, which come from their critique of old values. You must try your best to open your thoughts and accustom yourself to doubting old concepts or "truths". However, this is not easy to achieve. When we are confronted with a task, old ideas and opinions will naturally emerge as a sort of experience. You must constantly warn yourself to break away from them, make continuous attempts and ask critical questions. The key point of the project of "nothing is deeper than a surface" is the understanding of the concept of "surface". The inertia of most people's thoughts locates the design around the schemas presented by the outer skin of objects. However, when you look at the surface from a completely different perspective, and get to know the surface as a fragment in the process of the development of things, you will discover that the things hidden beneath the surface can reflect the depth and process of things all the more. The works of Ji Xin and Zhang Xiang express the deep reflections of the designers: "Is the surface the outer

开放，甚至产生新的意义。不断积累的丰富思想财富使我们的心智更趋成熟，当你解决问题时，从中选取的的潜在指示就越多，简而言之，你可以通过一个根本不同的方式来创造新的机制。

总之，人们通过知觉认识世界、改造世界、使一切设计行为得以展开。知觉一旦以经验主义的方式被定义为我们保存在脑海中对事物性质的一种占有就形成了心智。我们对于科学、对于艺术、对于设计的理解其实正是基于自身的主观意识对世界探索与体验后的结果，设计师将这种体验转化为各种具体的形态和图式向他人展示与表达。因此，我们可以说，感知决定心智从而控制了最终的设计图式；而图式也成为一种符号，向我们透露了设计者的心理意识。

通常，好的设计作品总是试图用最适当的图式来解释设计观点。但是，什么样的形式才是最恰当的形式呢？每个人的选择都有可能不同，但是往往只有那些看似简单，却恰如其分地表达出丰富含义的形式才最为打动人心。因为这符合了人们在知觉意义上把握事物特征的极简原则。最令人满意的简洁形式意味着最小的能耗而非奢侈与浮华，把一个观点讲得过于复杂与把它讲得过于简单一样是一件糟糕的事。在此次工作营的最终成果中，许多学生作品很好的把握了作品简洁性与复杂内涵之间的平衡。比如钱峥与马广超的视频作品"生命开始于结束"诠释了一个具有深刻涵义的主题：任何事物在经历了或长或短的生命周期后终将走向灭亡，从这个意义上来看，所有事物的表面就算如何的多姿多彩、与众不同，在开始的时候就注定走向同一个终点，这里"表面的深度"被理解成为一个生命过程。两个设计者采用了非常直观的具有诗意的形象仅在一分多钟的时间内便展示出这一作品的观点，所采用的中国画形式充分利用了该材料的物理机能，作品形式洗练、主题深刻、富有诗意。可以说，这个作品是复杂的，但在复杂之上显示出的简约风格显示了作者的心智在把握作品形式与主题关系上的成熟。

随着人们对世界的探索与体验的不断深入，事物也将不断地被知觉重新把握、重新确定；而心智的逐步完善与成熟将导致设计师们随时随地准备用新的图式来重新定义他们对世界的理解。在这种前题下，设计教师的工作也将被重新定义。教师不应再是灌输教条或是将自己的经验强行灌输的人，他们应该扩大学生的兴趣范围，唤醒学生的热情、感受能力和好奇心。给予时间，提出问题，促使学生自己思考和探索。对于建筑学学生，课题的设置不应只局限于围绕建筑的方面展开，而应扩大与此相关联的圈子，比如文学、音乐、数学、生物、舞蹈、绘画等等，使他们更早的打开精神空间，将新的事物引入他们的领域。正如此次联合教学的所展现的那样，一百多个学生以一百多种不同的图式来回答和定义他们对同一个课题的不同理解。这一过程他们所经历的、所痛苦的、所获得的，使他们的整个大学生涯甚至是日后的设计工作都将受益非浅。

skin?"; "Is the surface real?" The first work regarded a cross-section as a surface, so as to reflect the internal structure and texture of things. This texture has a close relationship with the sectioned object. The expression of a cross-section can indeed reflect the essence of things. The second product ("the deceptive surface") further reveals the confusion inherent in surfaces. Sometimes a surface is mere decoration: it cannot reflect the essence of the thing, and its internal structure is yet to be revealed. The work aimed to display the impact of the process of revelation. The techniques used in a "crude" way highlight the contrast between surface and essence. Here, what the product reveals is not a simple result, but the unique point of view and thinking process of the artist regarding this question, expressed by means of schema. Knowledge makes people tend to search for a recognizable part, because this part can be quickly interpreted and understood. However, if it stays at this step, the psyche has no chance to go further. The more you doubt inherent meanings, the more easily you will carry out analysis. In this way, your thoughts will become more active and open, and even give rise to new meanings. A rich amount of ideological resources accumulated continuously makes our psyche more and more mature, and you will be able to select more potential instructions from it when trying to solve a problem. In short, you can use a variety of different methods to create connections.

In a word, people get to know and transform the world through consciousness. In this way all acts of design are unfolded. Once consciousness is defined empirically as a sort of occupation of the natures of things that we keep in our mind, psyche will come into being. Our understanding of science, art and design is in fact based on the result of our exploration and experience of the world through our own subjective consciousness. Designers transform this sort of experience into various concrete forms and schemas and reveal and express them to others. Therefore, we can say that perception decides the psyche and controls the final schema of design. As a kind of tag, schemas also disclose to us the psychological constitution of designers.

In general, a good work of design attempts to explain an idea with the most appropriate schema. However, what kind of schema is the most appropriate? Different persons may have difference preferences. However, usually those forms that seem simple but can appropriately express rich connotations are the most touching, because this conforms to the principle of extreme simplicity, making it easy for people to grasp the characteristics of things in consciousness. The most satisfactory form means the most efficient one, rather than the most luxurious or ostentatious. At the same time, explaining an opinion in too complicated a way is no better than explaining it in too simple a way. In the final results of this camp's exercises, many students' works grasped the balance between conciseness and complicated connotation very well. For example, the video of Qian Zheng and Ma Guangchao named "Life Begins from the End" interprets a theme with a profound connotation: After experiencing a life cycle, long or short, all things will eventually die out. In this sense, no matter how varied and unique the surface of things appears, they are destined at the very beginning to go to the same destination. Here, "the depth of surface" is understood as a life process. The two designers execute the idea of the work within only one minute by adopting very vivid and poetic images. The form of Chinese painting adopted by them makes full use of the physical merits of this medium. This work has a polished form, a thorough theme, and a rich poetic flavor. It is complicated, but the concise style of execution on the complicated foundation reveals the maturity of the artists' psyches in grasping the relationship between the form and theme of the product.

With the continual intensification of people's exploration and experience of the world, things will also continually be re-grasped and re-determined by consciousness. The gradual perfection and maturity of psyche will induce designers to redefine their understanding of the world with new schemas whenever and wherever possible. On this premise, the work of design teachers will also be redefined. Teachers should not forcefully inculcate doctrines or their own experience. Instead, they should try to expand students' scope of interests, and awaken students' enthusiasm, sensation and curiosity. We should offer enough time and raise questions for students to ponder and explore on their own. With regard to students of architecture, projects should not be centered only around architecture; we should expand its scope and increase its relations to other disciplines, e.g. literature, music, mathematics, biology, dance and painting, so that the disciplines can open their spiritual space earlier, and welcome new things into their fields. Just like what was revealed in this joint instruction, more than 100 students completed and defined their different understandings of the same project with more than 100 different schemas. What they experienced, endured and obtained in this process will greatly benefit their entire life career at college and even their future design work.

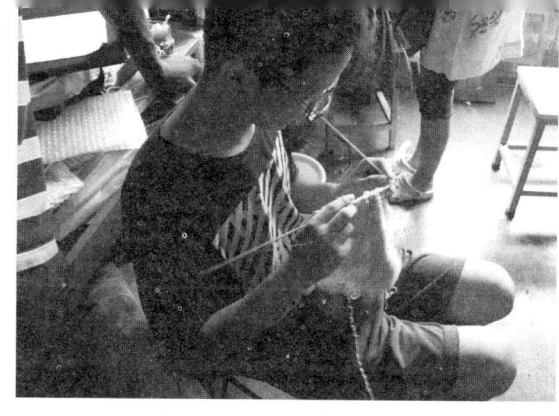

横看成岭侧成峰
——从理解差异到万物有源

胡碧琳

2010年的暑期后短学期,有些与众不同。

以绘画·空间·设计为主题的中法联合教学项目给酷热的8月吹来了一些新鲜的异域之风。有着不同的地域文化背景和办学历程的法国巴黎玛拉盖国立高等建筑学院的法国Philippe Guerin(葛汉先生)带来的系列课题,是从艺术史的角度开始,以空间及设计为延续的对学生进行设计思维的专门训练的渐进式课题,对建筑系二年级的学生,对传统的短学期课程,它无疑是新鲜自由的。

之所以新鲜,是因为之前艺术基础的实习课大多侧重于传统教学模式下素描、色彩的技法训练,而这次中法联合教学的创作课题在思维方式上焕然一新,不限制创作思维模式,不强调某种表现技法,强调的多是理念、思维过程的发展表达,是一种基础而宽泛的思维训练。例如:课题一(课题一:根据引言的创作"没有什么比一个表面更有深度")——葛汉先生通过展示西方艺术大师从乔托(Giotto)到马蒂斯(Matisse)的若干作品图片及评述,来说明了文艺复兴之前的表现形式和现代主义的表现形式是如何演变联系起来的,从绘画作品到装置艺术品再到影像作品的分析。这样的讲述教给学生从不同视角、不同时空去研究某些相似的东西,对学生发散性思维的训练起到良好的作用。于是,同学们有的从哲学思辨的角度来挖掘真实平面与其心理深度;有的从二维表面与三维空间上的深度的对比入手;有的从平面深度的宏观与微观比较表达入手;有的从虚实的角度入手,探究其体积感与延伸感……于是,他们发现平面(表面)—深度这对矛盾现象中,平面(表面)是可以被干涉的,是可以影响深度的——

"原来表面自有深度,因为表面总是通过距离、光影或空间达到深度的本质"(翟炼)

"一个平面可以有时间的深度"(原满)

"表面的深度也可体现在表面的可进入性和穿越性"(刘琦)

"表面的深度体现在人脑的思考深度"(晏莉莉)

"视觉的变换也可能改变表面的深度"(晨笛)

"不同介质下,表面会表现出不同特征,尤其在光的帮助下……"(王锁文)

这些"发现"表明了同学们的思维正慢慢从线性地走向多维的网状立体的思维。

之所以自由,是因为直接进入创作的课题会牵引出众多的想法、实现

Different Perspectives Lead to Different Results
—From understanding differences to the origin of universe

Hu Bilin

The short semester after the Summer of 2010 was a bit different than usual.

In the sweltering hot August weather, the Chinese-French joint teaching program centered around drawing, space and design brought an exotic freshness. Philippe Guerin, from the Ecole Nationale Supérieure d'Architecture Paris Malaquais, brought the French style of running a class. The series of tasks he prepared began from the perspective of art history and extended into space and design. Each training exercise built on the last and was specially designed to develop students' capacity for thinking about design. From the perspective of the second-year architecture students as well as that of the traditional short semester curriculum, this class was without a doubt novel and unrestrained.

What made this year so novel was that the Arts Education Internship class usually relies on the traditional method of training skills like sketching and color techniques, but this year's joint teaching with France had a completely new look. There were no limits placed on creative thinking, and no emphasized method of expression. Rather, ideas were emphasized along with achieving a developed expression of the thought process. It was a fundamental and wide-ranging exercise in thinking. For example, the first task (of three) was based on the introductory statement, "there is nothing deeper than a surface." Mr. Guerin lectured from Giotto to Matisse while showing clips of various works from masters of western art and their corresponding critiques, and explained how the representations in pre-Renaissance works are connected with those in Modernist works, including analysis of different works from paintings, to installations, to portraits. This narration allowed students to study similar objects from different perspectives, different places and times, thereby developing their skills of divergent thinking. Students dug beneath surface truth into more psychological depths with philosophical speculations: some proceeded from comparing 2 and 3-dimensional space; others began from a comparison of macroscopic and microscopic methods of expression on a plane; others proceeded from the perspective of falsehood versus reality, exploring the qualities of volume and extension… In so doing, they all discovered that in the contradictory pair of plane (surface) and depth, the plane can be manipulated. It can influence depth.

"It turns out all surfaces have depth, because they can use their distance, shadow or space to achieve the essence of depth." (Zhai Lian)

"A plane can have depth in time" (Yuan Man)

"The depth of a surface can manifest itself in its ability to be entered or penetrated." (Liu Qi)

"The depth of a surface comes from the depth of thought in the human brain." (Yan Lili)

"Changes in perception can change the depth of a surface." (Chen Di)

操作方法及众多的效果。在这些"众多"中,要靠积极行动和坚持己见,要在诱惑中放弃一些、选择一些,所以这种自由也被同学们认为是有"太自由的压力"。

有了这么多想法和认识,创意忽闪忽现,概念及观点得以自由的游走,自由到钻文字牛角尖,自由到表现跟不上想法,自由到偏执于矛盾的两极。同学们在接到第二个课题("所有的建筑都是对运动的框景""线条的图像、图像的线条"根据对引言的理解,结合前面讨论及习作的研究成果,不限材料及手法,对之前研究作深入或跨越式的发展)之后,忽然有些自由得不知所措……题目是由法文译成中文的,两种文字在原本理解和翻译之间可能存在差异,加上每人对这相对抽象的话的多种解释使大家的思维飘忽不定。

葛汉先生接下来充实又重要的第二个讨论十分及时。

这个讨论主旨是引导学生对艺术作品以及20世纪以来的变化有一个深刻思考,同时要找寻找自我定位,或者以理论的角度来质询历史性、现代性和当代性的作品。

正是这里,出现了些真空的断层。如果同学们对于一些与建筑设计息息相关的,如绘画、雕塑、哲学、美学之类的学科学习的不够全面,不能融汇贯通,这个讨论与之后的思考就只能是表层的,对现代、当代作品的学习和基本取样也随之浅显。不过,这又是一个极好的提醒和开始,因为大家还有时间和机会去学习。跨学科的学习会让大家拥有新的体验和开启新的思维方式,进而促进各学科的思维理念加快融合。和谐的、整体的、多元的、网状性的思维创造就来源于学科间的交叉创新。

第二个课题有的学生浅尝辄止,

有的学生挖掘到了深层的东西。

有的以综合生理及心理的体验来诠释和再现物像的虚实,正如题目中运动对于固定的建筑的虚实转化亦或点线面之间的虚实转化;有的以抽象的轨迹记录生命时间、社会空间、地域文化空间并混合现实与抽象;有的用平面方式呈现解构的空间。同学们渐渐开始用哲学的分析,从自己个人的世界观入手来进入创作了。

通过葛汉先生两次针对现代、当代纯艺术作品的分析讨论、作业讲评、个别交谈及分组讨论,第三个自由命题的创作大家已不像第二课题那样把握得飘忽不定了,对于前两个引言的分析,大家已有了不同角度的视点和结论了。有的沿着既有线索独立

"In different mediums a surface will show different characteristics, especially with the help of light…" (Wang Suowen)

These "discoveries" demonstrate how students' thinking in the process of moving from linear thought to a 3-dimensional network of thought.

This type of education was unrestrained because directly jumping into creative tasks dug out a multitude of ideas, methods to realize them, and a multitude of results. In all these multitudes, one must stick to positive action and preserve one's subjectivity. When tempted by a new idea, one must be willing to sacrifice some things and to choose others. As a result, students felt that this kind of freedom "gave freedom to stress."

With so many thoughts and ideas, creativity flickered. Concepts and points of view flowed freely into the deepest depths of language itself, so freely that expression couldn't keep up with the pace of thought, so freely that one would get stuck between the two poles of a contradiction. After the students received the second task ("All buildings are windows into movement", "the picture of a line, the line of a picture"… Based on your understanding of the introduction, combine the preceding discussions and the results of your first exercise to go deeper into the topic or develop it further by connecting it with another. There is no limit on material or technique.), students were suddenly encountered with so much freedom they were at a loss. The original topic was written in French and then translated into Chinese, so perhaps some content was lost in translation. In addition, the abstract nature of the statements led each individual to their own unique interpretation, making students feel as if they were floating alone in chaos.

Mr. Guerin proceeded to substantiate things with his second discussion, very timely.

The main function of this discussion was to lead students deeper into thinking about works of art and post-20th century art, at the same time searching for one's place in the larger scheme of art, or to inquire into historical, modern and contemporary works of art through a theoretical perspective.

It was precisely here that students slipped into a void. If students lacked a comprehensive understanding of the subjects related to architectural design, like drawing, sculpting, philosophy, aesthetics, etc., they couldn't speak with authority and the discussion and subsequent speculations would only scratch the surface. Their survey of modern and contemporary works of art would be similarly shallow. However, this was also a tremendous awakening and beginning, as everyone still had the opportunity to continue studying. Multidisciplinary studies could give people new experiences and activate new styles of thinking, while facilitating the study of each subject and the synthesis of the respective thought processes. A harmonious, holistic, multi-faceted, and interrelated creative thought process came from innovation at the boundaries of disciplines.

After the second task, some students had just begun to scratch the surface, while others were busy digging up deeper discoveries.

Some experiments in physiology and psychology represented and explained the true nature of things, like the projection of movement onto static architecture in the second task's introductory statement, or the transference in the relationship between point and line. Some experiments used abstract paths to register the life of the times, the space of society, regional cultural space, with a mixture of reality and abstraction. Some used planes to deconstruct space. Students gradually began to use philosophical inquiry to use the world of their own experiences as a starting point for their work.

Having gone through two rounds of Mr. Guerin's analysis and discussion of fine art

而坚持地探索；有的发现了新的感兴趣的路线方向……

这时寻找合适、快捷、真我、精彩的表现方式是大家共同面临的新课题。教室里，大家忙于试验各种想到的材料和加工工艺，尝试多种表现媒介，通过摄影、绘画、影像、装置等方式表现思想。在一次次的尝试中，有时候结果不可预测，过去的现成经验与方法突然会变得不确切起来，由于结果不可预测，过程似乎变得危险起来。遭遇未知和失败使同学们发现了新的方式、新的现象，于是有了新的可能性，对同样的问题会有不同认识，不同的解决方案，不同的操作模式，表现方法以及不同的结果，这就是这段课程的价值和期待所在吧。

也许，我们同学因此学会了用不雷同的美去证实世界的丰富存在，会明白客观审美与个人体验的微妙关系将是未来创作的核心，"横看成岭侧成峰，远近高低各不同"，中国历代的先贤哲人早就用生活体验提炼了大千世界，从这一点来说，艺术创作的本质是不分国界的。

这也是本次课程带给大家最大的收获。

这样的思维训练课程教会了我们开放与接纳、质询与探究，学会理解创作过程中的形式与内容、本质与表象、共性与个性的辩证关系。课程总是提示起点、不限空间、也不追究结果，容许违反原理的东西出现，鼓励大家寻找隐藏的矛盾，让大家学会不是在非此即彼的关系中成长，而是在此与彼之间不断地发展新异。

短短两周半时间，同学们丰富精彩的作品，使我们深深认识到引导同学的个人灵感体验，鼓励个人的探究能力远比教会他们技巧方法更为重要。

我们在教学过程中虽有事先确立的目标，但过程并不一定按预想的方向发展，其间充满未知，充满乐趣，充满各种创造可能。没有现成答案和唯一标准，只有充满活力的对话与讨论，探索与碰撞，在与同学的头脑风暴中，大家只尽力延展思想使其不至于僵化，这也考验了老师们的应变判断与知识结构，也促使大家去不断研究学习，解决新问题，涉足新知识领域。

from modern and contemporary periods, his critique of their coursework, and countless conversations and group discussions, students felt like they had a more firm footing going into the third task. Through analysis of the previous two introductory statements, students had formed their own conclusions from a variety of angles, with a variety of viewpoints, some following the threads provided by the course and initiating independent discovery, others discovering their own new and interesting avenues of inquiry.

This time, everyone's mutual goal was to find appropriate, elegant, true-to-self, and exciting methods of expression. In the classroom, everyone was busily experimenting with all kinds of ideas of new materials and techniques for applying them. Students tried new media like photography, drawing, silhouette, installations, and more, in order to express their ideas. In the countless experiments, some of the results were difficult to predict, as the standard set of experiences and methods would suddenly become indefinite. As a result, the process sometimes was exciting and even dangerous: Running up against the unknown and even failure made the students discover new methods and new phenomena, thereby grasping new possibilities. Facing a problem, students had their own understanding, their own solutions, their own methods of operation and expression, and unique results, perhaps just where the value of this course was expected to be.

Perhaps it was because of this that we students learned to pave our own way and use a unique aesthetic to prove the richness of the world, and learned that the subtle relationship between objective aesthetics and individual experience is the core of creation. "It's like a range when viewed from the front. But it's like a peak when viewed from the side. The mountain showed its different features when viewed from different levels near and far." Ancient Chinese scholars had long ago captured the universe with philosophical insight through life experience. From this perspective, the essence of artistic creation knew no national boundaries.

This was everyone's biggest lesson of the class.

This course trained our thinking abilities and taught us to open up and accept, to inquire and explore, to understand how the creative process is composed of form and content, essence and representation, and both common and individual ideas in a dialectic process. The constant reference to a starting point, an unlimited space, the option to disregard the results, and the acceptance of things that contradicted principles, all encouraged everyone to seek out hidden contradictions. We learned that growth doesn't happen in black and white, but that somewhere in between there is an endless source of novelty.

In a little more than two weeks, the student's rich and exciting works made it clear that experiencing inspiration and encouraging exploration was far more important than learning methods and techniques.

Although we had a pre-established goal for the course, things do not always progress as expected. The period was full of unknown, full of interest, and of possibilities for creation. Without pre-set answers and material standards, there was only lively dialogue and discussion, the exploring and colliding of ideas. With the students' minds in a thunderstorm of activity, everyone did their best to extend their lines of thinking and prevent them from becoming rigid. This process also tested the teachers' abilities of quick thinking and framing knowledge, and let everyone continue to make studying an object of study, solving problems and setting foot on new territories of understanding.

关于加强美术教育在设计课程中功能的思考
——从作为教学补充与改革的联合教学想起

戴斐

以审美教育，或者狭义地说，以美术教育修养身心、提高人生质量，并非是一种始于当下或舶来的时髦话题，而是有着深刻历史底蕴的文化识见，历经近代以来的几番反复方得践行。作为审美教育主要内容与途径的一门非专业艺术教育，美术教育有着很强的针对性，而不限于通常理解的"人文素质教育"范畴。

在建筑设计相关专业的教学中，美术教育素来是易被忽视的环节。不仅许多学生冲着"画好CAD图就能找到好工作"来选择专业，甚至在一些专业教师的言语观念中，也常常会自觉或不自觉地流露出重工偏文、技大于艺的倾向。这样一来，原本只是人手替代物的电脑设备被人为地凌驾于人脑之上，大有取代人脑的聪明才智和创造性思维的趋向，这种急功近利的教育模式的结果，用通俗的话来形容就是：造成一代有知识没文化的快餐型人才的产生。

正因如此，本次中法双方联合教学作为一次与之迥异的、新兴教学模式的输入与尝试，尤显可贵。它的意义在于秉持了一种有异于常规智力性、技能性教育的培养观念，在课程讲授上以训练学生的思维能力与表现手法为主，"并非在于传授学生某种设计或创作的具体技法，而是力图让学生在了解西方艺术及其历史的同时，学会发展自己的艺术观，培养自己对作品的独立思考能力和批判性精神。同时，在此基础上，对于在当今文化日益全球化的情境下，对自己传统文化的走向有一个独立、深入的思考。"以艺术史的贯通来形成个性化的观念并运用于设计，类似教学方法在现今设计专业尤其是工科专业中尚不多见，但这类方法对全面知识体系的形成和人文精神的提升大有裨益，也是创新和独立思考能力培养的基础，而这些，恰恰正是常规智力教育的薄弱之处。

由于涉及现行教育体制的革新和众多学科成果的支持，此类颇具创新意味的美术课教改只是小范围的浅尝辄止。以观念为主导的美术教育真正在现行教育方针中的实施以及相关理论的延伸，并非一蹴而就的易事，需要一个长期的过程。其内容及必要性主要基于以下几点来进行：

1. 完善知识结构，培养基本审美能力

当代美育学科理论体系曾就"美术教育"一词给出了明确定性，将其定位于：从属于教育学、介于多门学

On Strengthening the Role of Arts Education in Curriculum Design
—From the perspective of joint teaching functioning as a way to augment and reform teaching

Dai Fei

Education in aesthetics, or more narrowly speaking, in the fine arts to cultivate the mind and body and raise one's quality of life is not something that emerges in the present age, nor is it some exotic, trendy topic. On the contrary, it embodies deep historical and cultural insight, and has undergone repeated changes since the beginning of the modern age before it can be actually implemented. Comprising the main content and methods of aesthetics education, general fine arts education is extremely relevant, and certainly not limited in scope to a "humanistic education," even though it is commonly understood this way.

Among education related to architectural design majors, fine arts education has always been an easily-overlooked element. Not only do many students choose the major because they believe that simply learning to use CAD (computer aided design) software is enough to guarantee finding a job, but even the rhetoric of some instructors often consciously or unconsciously reveals a bias toward engineering over the humanities, or toward technique over art. In this way, the computer, which was once merely an alternative to the hand, is now being elevated above the human mind and quite possibly may replace human wisdom and creative thinking abilities. The resulting method of education that focuses so much on instant benefits, to put it bluntly, has created a generation of fast-food style talent production, creating professionals with knowledge, but lacking culture.

Precisely because of this situation, as it represents the introduction of a new burgeoning style of teaching, and an experiment in something widely different from the norm in the field, our Chinese-French joint-teaching project is of the utmost value. Its significance is in its upholding a philosophy of training that departs from technique-centered conventional wisdom. During class lectures, the main goal was to develop students' thinking abilities and techniques of expression. "It was not to impart with students any certain specific technique of design or creation. Instead, while seeking to help students understand western art and its history, it encouraged students to develop their own unique conception of artistic works and cultivate an independent thinking ability and critical spirit. Moreover, in the midst of the present situation of the continuous internationalization of culture, it sought to help them move towards an independent and deeper reflection on their own traditional culture". Nowadays, educational methods that emphasize a thorough understanding of art history in order to enable the creation of individual ideas and their use in design are seldom seen in the design major, and even less so in the engineering track. However, these methods have immense value for forming a comprehensive knowledge system and advancing a humanist spirit. They are

科之间、包容丰富的交叉学科。这一定义客观上对全面而完整的知识结构提出了要求。个人知识体系的构成情况、结合方式与掌握程度形成了一个由诸多要素组合而成有层次、有序列的整体知识信息系统，反映了一个人的文化素质。现代美术教育理论的体系化建构摒弃了当代社会工业化和唯科学主义泛滥的弊端，拥有"知识经济时代不同于通常的工业经济时代的特征"，不仅要结合文化发展趋势，关注学科本身的历史延续性，更须在从方法到内容等各个方面，都与现代心理学、信息科学、生命科学、思维科学、脑科学等诸多相关自然学科成果进行联姻，这些学科自身的不断进展和整合研究能够为美术教育与时俱进的发展提供必要的学理依据。

对于设计专业而言，美术教育的作用突出体现在独特的构思，高超的审美力上，而这些与丰富的工程知识、娴熟的技术能力一样，都是构建与完善设计师素质不可或缺的因素，因在文化精神形成过程中所起的打底作用，有时甚至更为重要。作为造型艺术之一的建筑设计学科是以综合性美术知识为基础，其艺术形象的建构涵盖了包括比例之美、色彩肌理之美、环境之美、空间分割与组合之美、人文与自然之美等在内的众多内容。涉及的学科范围也不仅仅限于美术，而要宽泛得多，涉及到结构力学、建筑材料学、施工与工艺学、人体工程学与心理学等多领域知识。一名优秀的设计师应该是集科学技术与艺术素养为一身，拥有较为整体的知识结构的全才，需要多年的文化浸润方得厚积薄发，绝非在校期间屈指可数的必修课和短短数年专业强化学习可以达成。

对美术教育课题的研究立足于时代特征和人的基本审美力，这是人们借助形象的一种特殊情感判断能力，也是健全人的心理结构和智力结构中不可或缺的元素。把审美力问题作为美术教育基点，将理论的深化与人在当代文化语境和人性发展层面上的精神状态联系起来，这一研究方法向上承袭了十九世纪初期以来的美学理论线索，向下直面当代人性发展所遭遇的失衡与冲突问题，通过促进人的丰富想象力，协调人在现实中的情感心理，形成创造美好生活的愿望与能力，实现生命的和谐发展，所谓"现代人文价值"即在此。

2. 训练本体技能，锻炼创造运用能力

美术作为一种具体的艺术实践活动和艺术表现形式，是一个完整、自足的循环体系，有其自身发展规律，因此具有内在的、自然的发展动力，即作用于美术自身的本体功能。西哲认为"所谓本体，指终极的存在，也就是表示事物内部根本属性、质的规定性和本源。"一门课程之存在价值，是由它无可取代的独特知识体系，技能技巧和教育功能的特殊性决定的，这种区别于其他学科的完整的自身结构和价值意义便构成了课程的学科本体。美术学科的本体属性特征界定了教学开展的范围，强调了学科的人文性质，并明确表明其教学必须从严格的技能训练入手，进而感悟精神内涵。因此即使是"并非在于传授学生某种设计或创作的具体技法"也不是否定知识技能，而恰恰是在人文

the foundation of creative and independent thinking abilities, but remain the very weak point of conventional educational wisdom. Because they are connected to the problems of reforming the current educational system and of supporting achievements in a wide range of courses, these kinds of creative methods of fine arts educational reform are a small-scale experiment. The implementation of fine arts education under current educational policies and the extension of theories related to such fine arts education can not be easily accomplished. It will be a long-term process. The content and necessity of fine arts education are illustrated in the following points:

1. Optimize the structure of knowledge; cultivate fundamental aesthetic abilities

The system of contemporary aesthetic education theory has clearly defined "fine arts education" as: a content-rich interdisciplinary study which is a part of educational studies and exists at the intersection of many different disciplines. This definition obviously calls for a comprehensive and complete structure of knowledge. The composition of one's knowledge structure, one's methods of connecting things, and the degree of one's grasp of things form a holistic system of knowledge and information composed of multiple ordered elements at different levels—this reflects one's level of culture. The systematic structure of current fine arts educational theory, however, has abandoned contemporary society's industrialization and the drawbacks of rampant scientism. It possesses the "the qualities of the economy of knowledge as they differ from those of the usual industrial economy". Not only should it integrate the development patterns of culture and consider the historical continuity of science itself, but, even more importantly, should make connections ranging from methodology to content between the achievements of modern psychology, information science, life sciences, the science of thinking, neurology, and other natural science disciplines. The continuous evolution and integrated research of these disciplines can provide a theoretical basis necessary for fine arts education to keep up with the changing times.

As for the design major, the effect of fine arts education primarily shows itself in producing unique abilities of conceptualizing and finding beauty. Along with rich engineering knowledge and skilled technique, these qualities are indispensable to a perfect designer. Their grounding effect in the process of forming cultural spirit is sometimes even more important. The discipline of architectural design as a visual art takes artistic knowledge as its foundation. Its artistic essence is constructed of elements covering the aesthetics of proportion, color texture, environment, spatial division and recombination, philosophy and nature, and many more. Its range of disciplines is much broader than merely being limited to the fine arts, involving many areas including the studies of structural mechanics, building materials, construction techniques, ergonomics and psychology. An excellent designer should possess both technological ability and artistic sensibility and have talent spanning a comprehensive range of knowledge. This requires a steady accumulation of many years of cultural experience. It is not a status that can be attained during the few required courses and a few years of intensified professional learning in a student's college career.

Our understanding of fine arts education and the research of its related topics is established through both the spirit of the times and through people's fundamental aesthetic capacity. It is a particular judging ability driven by emotion that

精神统领下，依仗学科本体内作为视觉艺术的绘画、雕塑、工艺等最有特征、最基础的知识技能来构建教学本体并综合实施的。

3. 拓展创新思维，引导个性化观念形成

每个个体都具有强大的创造潜力，这种没有固定格局的创造力是与生俱来的。以多元智能理论的观点，每个人都拥有多种在人体中分布不均匀、显著的智能，并以其特有方式组合而形成个体独特的智能结构，体现在人的能力和个性上就呈现出五彩斑斓、争奇斗艳的效果。个体智能本身的多样性就决定了必须实施个性化教育，然而实际社会所给与的墨守成规思维强加于这种能力，强制将独具一格的思维统一为固有的套路。针对此类应试教育型教育弊病而生的对于教育体制革新的思考，则是在有关个人评估方式的研究上，把属于"一种非智力的情感领域"的、根本目的在于"确定一种审美的态度和人生观"的现代美术教育纳入评估标准。教学的过程，就是一个立足于学生智能与个性的差异，尊重每个学生不同的能力、兴趣、气质和性格特征，鼓励其发挥积极性和主动性，帮助其形成完整、健康、充实的个性特征的过程。学生们通过联合教学的发散性思路教育，尝试按照自己的意愿做出自己想要的东西，没有应该做的，没有必须做的，有的只是按自己对于艺术的理解，用自己的视角诠释这个世界。而这些是他们在惯常标准的、科学的设计专业课的教学大纲中找不着，学不到的。

4. 贯通史论，以动态、整体、发展眼光审视艺术现象

美术史中涉及的和美术相关的文物及文献以及时代流变过程，对于现今了解古代时期的政治、社会、经济、文化的发展有着重要的作用，有助于研究者形成一种历史反思与现实关怀、学理探究与实践追求相结合的思想风格。这种以史的观念和视角来认识与比较、用通贯古今、发展性地看待问题的学习方式，可以锻炼学生使之逐渐发展起批判性的思想，自觉地参与到一种自我质询与辩论中，一方面以理论的角度来质询历史性、现代性和当代性，另一方面又以实践的角度来审视传统的手法和革新的技法，特别是思考关于国际化下的艺术现状以及它与当代世界关系的演变。将原本作为个案的某个艺术设计置身于一个史的广阔范围进行涤荡、考验，为其准备一个与历史相关联的艺术定位，以此眼光来看待设计行为及其作品，可以在设计中扭转短期化、狭隘化、功利化的设计心胸，训练设计师的宏观气魄。

当代的生存境遇是发展与破坏并行，前进与倒退共存，处在一种窘迫的"非美"状态，也是生产力与人性关系发展到一定时期遭遇的"瓶颈"。在现时代条件下，进行美术教育应以培养"生活的艺术家"为己任，从关注人的生存状态，使其得到有效的改善和提升出发，尽快将美术由古典形态的对美的抽象思辨这个窠臼中脱离出来，实现对美与人生关系的探索这一美育转向，由哲学美学转到人生美学，以审美的心态和行为对待社会和自然，尤其是人本身，做到

we have in the presence of images, which is an indispensible element in the sound construction of a psychological and intellectual constitution. This method regards aesthetic capacity as the foundation of fine arts education, and connects the deepening of theories and people's metal status under contemporary cultural context and at the level of the development of human nature. This method follows the aesthetic tradition which started at the beginning of the 19th century and squarely faces the problems of the loss of equilibrium and conflicts arising from the development of human nature in modern times. It makes aesthetic ability the foundation of fine arts education and connects the development of educational theory with the spirit of contemporary culture and human development. By enriching people's imagination and tempering their emotional psychology, it engenders a desire and ability to realize a beautiful life and to achieve harmonious development. It embodies "modern human value".

2. Train noumenal skills; forge abilities of creation

As a concrete artistic practice and form of expression, the fine arts make up a complete, self-contained system. They have their own law of development, and hence have naturally inherent momentum, which is noumenal functions acting on the fine arts themselves. As in western philosophy, "The so-called noumenon refers to ultimate existence, i.e. the regularity and the origin indicating a thing's inherent fundamental nature and quality." The value of a course of study is decided by its unique and irreplaceable system of knowledge, its related skills and techniques, and its special educational function. Its inherent structure and value differing from those of other disciplines constitutes its noumenon. The inherent essential qualities of a course in the fine arts define the scope of teaching to be carried out, emphasizing the humanistic character of the subject and clearly dictating the necessity of proceeding from strict technical training to enable experiencing its spiritual content. Accordingly, even if it is "absolutely not seeking to impart students some particular design or technical skill," it is still does not deny technical knowledge. On the contrary, it is precisely under the command of the humanistic spirit, relying on the most representative and fundamental knowledge and skills of such visual arts as drawing and painting, sculpting and craftwork within the discipline itself to construct its noumenon of teaching and realize its overall development.

3. Expand creative abilities; guide the construction of individual opinions

Each individual possesses great creative potential which is without a prearranged pattern, and is inherent in the individual. According to the theory of multiple intelligences, every person has many outstanding aptitudes which are unevenly distributed. Each person uses one's own method of combining the abilities to form an individual intellectual structure which shows itself in the colorful possibilities in people's talents and personalities and the grueling competition which results between them. The very multiplicity of every individual's intellect determines that individualized education must be implemented. However, the conventional thought structure imparted by society is imposed on these abilities, coercing one-of-a-kind thought structures to congeal into a fixed set of abilities. Countering this kind of exam-oriented educational malady, thought that seeks to renew the educational system is based on the critical standards of each individual. It

人与自然、生理和心理的和谐发展。是当今教育所应具有的策略。贯穿于20世纪西方现代历程的人文主义美学思潮，从某种意义上讲就是一种人生美学，也就是广义上的美术教育。这一命题在人的自身发展问题上沟通了科学与人文，也正是对时代现实的一种适应。

brings modern fine arts education, "a non-intellectual emotional territory" whose fundamental goal is to "define an aesthetic attitude and life philosophy", into the standard of critique. The teaching process is a process that helps each student to form a complete, healthy and rich individual personality by basing itself on the difference between students' abilities and character, respecting students' different abilities, interests, disposition and personality traits, and encouraging students to release positive initiative, Through the divergent thinking instilled by the educational methods of this year's joint teaching project, students tried to create something based on their own desires. There is no "should do" or "must do"; there are only their own unique perspectives which can be combined with their understanding of art to explain the world. That is something that cannot be found and learnt in the standard, overly-scientific course outlines of the design major.

4. Acquire thorough historical knowledge; use dynamic, holistic, and innovative vision to view art

Cultural relics and literary contributions related to the history of art, as well as their changes over time, are effective nowadays for understanding the development of government, society, economy and culture of bygone ages. They help researchers form an attitude of reflection and appreciation of reality as well as an intellectual style that combines theoretical research with practice. This kind of educational method, which uses a historical perspective to understand and compare the old and the new, while expansively regarding problems, can train students to gradually develop critical thinking skills and consciously participate in self-inquiry and debate. On one hand they use a theoretical perspective to inquire into the historic, the modern, and the contemporary, and on the other hand they use a practical perspective to examine both traditional and innovative techniques, especially thinking about the evolution of the state of art under internationalization and the changes of its relationship with the contemporary world. Combing and testing an individual case of artistic design within a broad scope of history, determining an artistic positioning in relation to history, and viewing design activities and works from this persepctive can remedy short-term, narrow-minded, and utilitarian design aspirations and instill designers with a macroscopic, bold spirit.

Current existential circumstances find development and destruction hand-in-hand and the coexistence of progress and regression. We exist in a debilitated "anti-aesthetic" state, and the interaction between production and human nature is evolving to a roadblock. Under current circumstances, conducting fine arts education should cultivate the "life artist". By closely following people's modes of existence with a view to effectively improve and upgrade them, we can split the fine arts as quickly as possible from the set pattern of the classical mode of abstract inquiry into beauty, and then realize a change of direction in the relationship between human life and aesthetics and move from a philosophical aesthetics to a life aesthetics. Using an aesthetic state of mind and set of actions to approach society and nature, especially humanity itself, we can achieve harmonious development between humans and nature, between psychology and biology. This is a necessary strategy for current educational endeavors. Running through modern western history of humanist aesthetic thought, more or less is a kind of life aesthetics and a generalized aesthetic education. This proposition for one's personal development communicates a process of adaptation between science and humanities, in other words, an adaptation to

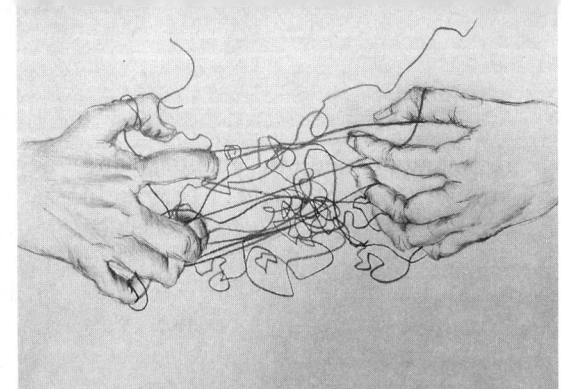

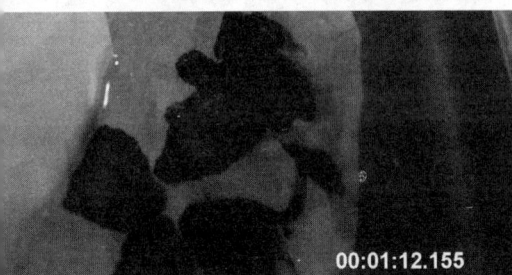

00:01:12.155

生命开始就注定会结束
如墨色四散开
或挥洒
或消亡

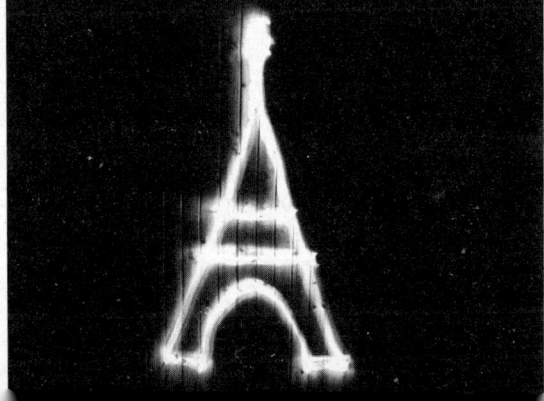

附录

视觉设计基础教学大纲

（总学分：4 总上课时数：231）
东南大学建筑学院

一、课程的性质与目的

视觉设计基础是建筑学、城市规划、艺术设计三个专业的共同学科基础课。它的教学目的是使学生通过本课程的学习掌握基本的造型能力和表现方法，提高学生运用视觉语言进行记录、表达和思考的能力，使学生既具有基本的造型能力，又掌握现代艺术的基本理念和表达方法，为三个专业的学习打下一个良好的视觉设计基础。

二、课程内容的教学要求（附录3）

1. 形体结构的分析与表现：几何形体结构与复杂物体结构的认知。
2. 视觉表现基础：石膏体表现的基本规律及表现方法。
3. 光影、材质研究：明暗构成规律、明暗等级的划分及质感表现。
4. 视觉分析与表达
   （1）空间研究：在透视环境中的空间认知与构成；
   （2）肌理分析：真实的、模拟的、抽象肌理的综合。
5. 综合手绘表现技法：
   （1）黑白线描写实规律的初步认识：静物的钢笔线描表现、风景的钢笔线描表现、建筑景观快速表达；
   （2）钢笔淡彩：彩色铅笔快速表现、马克笔快速表现、钢笔水彩快速表现。

6. 综合色彩构成表现：
   （1）色彩认知；
   （2）色彩分析；
   （3）色彩语言与运用。
7. 综合造型基础与视觉表现实验。
   （1）徒手表现训练；
   （2）视觉思维训练与综合造型实验或国际联合教学。

三、上机实习要求

无上机实习。

四、能力培养的要求

1. 分析能力的培养：通过本课程的学习提高学生对视觉现象、规律的认识、分析能力的培养达到对视觉形式的理解。
2. 自学能力的培养：通过本课程的教学培养和提高学生对所学知识的理解、整理、概括、消化、吸收的能力。
3. 表达能力的培养：通过不同阶段，不同作业的练习，表达自己解决问题的思路和步骤的能力。
4. 创新能力的培养：培养学生在视觉设计基础训练中的独立思考、深入钻研问题的习惯，提高解决问题的能力。

五、建议学时分配

见下页附表。

六、考核方式视觉思维训练

总评成绩＝平时成绩（作业）＋期末考试成绩；
平时成绩占40%；
期末考试成绩占60%。

七、教材及参考书

1. 顾大庆．设计与视知觉[M]．北京：中国建筑工业出版社，2002．
2. 梁蕴才，高祥生．钢笔画技法[M]．南京：东南大学出版社，1988．

1. 形体结构的分析与表现: 几何形体结构与复杂物体结构的认知。

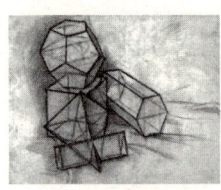
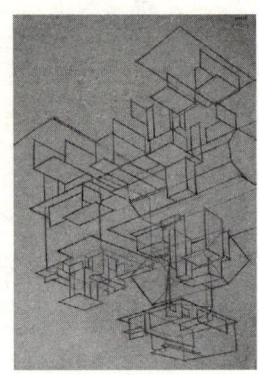

2. 视觉表现基础: 石膏体表现的基本规律及表现方法。

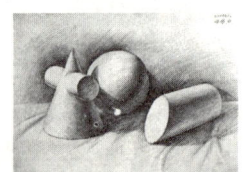

3. 光影、材质研究: 明暗构成规律、明暗等级的划分及质感表现。

课程内容	讲课	习题课或课堂讨论	实验	上机
形体结构的分析与表现	2	10		
视觉表现基础	2	10		
光影、材质研究		12		
视觉分析与表达		12		
空间研究		6		
肌理分析		6		
综合手绘表现技法		132		
钢笔手绘表现技法		48		
彩铅快速表现技法		24		
马克笔快速表现技法		24		
钢笔水彩快速表现技法		36		
色彩综合构成表现		12		
色彩认知	1	2		
色彩分析	1	2		
色彩语言与运用	1	2		
色彩的三维表达	1	2		
综合造型基础与视觉表现实验		39		
徒手表现训练		13		
综合造型实验或联合教学		26		

4. 视觉分析与表达
（1）空间研究：在透视环境中的空间认知与构成；
（2）肌理分析：真实的、模拟的、抽象肌理的综合。

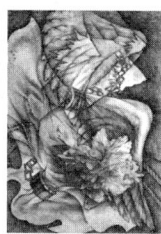

（2）钢笔淡彩：彩色铅笔快速表现、马克笔快速表现、钢笔水彩快速表现。

5. 综合手绘表现技法：
（1）黑白线描写实规律的初步认识：静物的钢笔线描表现、风景的钢笔线描表现、建筑景观快速表达；

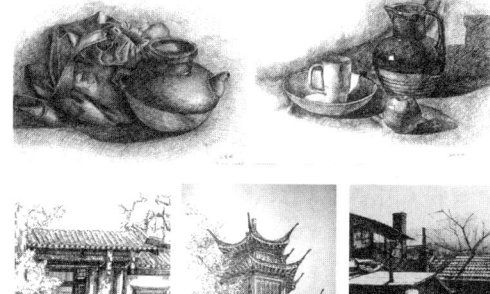

6. 综合色彩构成表现：
（1）色彩认知；
（2）色彩分析；
（3）色彩语言与运用。

7. 综合造型基础与视觉表现实验。
（1）徒手表现训练；
（2）视觉思维训练与综合造型实验或国际联合教学。

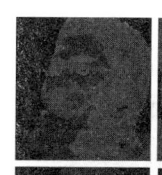

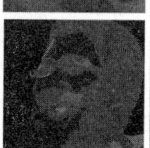

后 记

在总结这次活动之前，我想感谢所有直接或间接促成此次活动的人士。首先是东南大学建筑学院的老师和学生们，感谢你们满怀信念地参加此次教学活动和对出版所做的大量工作，同样也感谢几位法国同学以他们的方式在此融入了他们的学习体验。

其次，我想特别感谢龚恺先生对教学目标深远、开阔及清晰的洞察；感谢曾琼先生教学期间细致的陪伴和大力的配合；当然，还要感谢赵军先生从一开始大力推动了此次教学交流项目，并将其结合到了当今现实的（教学改革的）背景中。

最后，感谢宋磊先生和王盈先生，他们以主动的原创精神和细致的工作，建立了一个实在的、卓有成效的交流合作，使得我们所有人能够从中获益。

在我们这些天的工作中，充满了信任、好奇和互相聆听的气氛，这使得我们的文化及教学的交流变得可能，并得到鼓舞。整理、开放性的建议等，对于延续这一交流经验是非常必要的。同时，这本出版物首先是记录这次交流的具体的工具，并且它丰富了这一对话，以此便于我们开展更专注的项目。更确切的是，它可以对中国学生的中长期的学习历程有一个观察与分析。

首先，我们注意到中国学生对于技术的运用以及总体上很好的掌握，这使得我们好奇他们在将来的设计中，选择技术来发展他们方案时候的想法，并且他们自己是有着怎样的自由，敢于去发展他们的方案。

其次，当然也是我们中方的老师们根据中国建筑院校中的需求，来转化此次教学经验，我们也知道这种需求演变得很快，尤其是对于艺术和文化而言。

菲利普·葛汉

Epilogue

Before we conclude, I would like to address many thanks to all the people who made this project possible, either directly or indirectly. First and foremost, I would like to thank all the teachers and the students of your school of architecture who enthusiastically took part in the workshop and in the writing of this document, as well as the French students who contributed in their own way to a good integration of their own experiences.

Furthermore, I want to thank Mr. Gong Kai in particular for his strong, broad and clear point of view on the issues at stake, Mr. Zeng Qiong for his careful guidance and his ever dynamic availability, and of course, Mr. Zhao Jun for the impulse he gave to this project from the very beginning by placing it within a palpable reality.

Finally, gave my thanks to Mr. Song Lei and Mr. Wang Ying, who managed, through their spirit of initiative and their constant alertness, to lay the foundation for a real and valuable collaboration, from which we substantially benefited.

Epilogue

The atmosphere of trust, curiosity and mutual understanding in which we work today promotes and allows for the existence of this cultural and pedagogical exchange. Improvements and proposals for further thought will nonetheless be necessary for the experience to continue. Thus, this work is first and foremost a tangible tool which records the exchange and fuels the dialogue to allow us to develop our projects with an even greater degree of cooperation. It will also make the observation and the analysis of mid-term and long-term results of the Chinese students more relevant.
Firstly, our curiosity, which pertains to the techniques they use and master, will apply more precisely to the thought process surrounding their future works and the freedom they will allow themselves.
Secondly, it will be the task of our colleagues to transform these pedagogical methods accordingly with the needs of Chinese architecture schools — needless to say, these needs will be evolving very quickly, especially when it comes to art and culture.

Avant de conclure cet exposé, je voudrais remercier vivement toutes les personnes qui l'ont rendu possible, directement ou indirectement, à commencer par tous les enseignants et étudiants de votre Ecole d'Architecture qui ont participé avec conviction à ce workshop et à l'élaboration de ce document, sans oublier d'ailleurs les quelques étudiants français, qui ont contribué, de loin et à leur manière, à une bonne intégration de leurs expériences vécues.
Ensuite, Je tiens à remercier particulièrement Monsieur Gong Kai pour sa vision forte, large et claire des enjeux convoqués, Monsieur Zeng Qiong pour son accompagnement attentif et sa disponibilité toujours réactive, et bien sûr, Monsieur Zhao Jun pour l'impulsion qu'il a su donner à ce projet dès le départ en le positionnant dans une actualité tangible.
Enfin, pour terminer, tous mes remerciements à Messieurs Song Lei et Wang Ying, qui ont su par leur esprit d'initiative, doublé d'une vigilance constante, poser les bases d'une vraie et fructueuse collaboration dont nous avons tous pu largement profiter. Le climat de confiance, de curiosité et d'écoute réciproque dans lequel nous travaillons aujourd'hui, encourage et permet l'existence de ce double échange culturel et pédagogique. Des aménagements, des propositions d'ouverture seront nécessaires pour reconduire cette expérience. Aussi, cet ouvrage est avant tout un outil concret pour mémoriser cet échange et nourrir le dialogue afin de développer nos projets avec encore plus de concertation. Il rendra plus pertinente la possibilité d'une observation et d'une analyse des retours à moyen terme, et peut-être à long terme, sur les parcours des étudiants chinois.
Dans un premier temps, notre curiosité, certes attentive aux techniques utilisées et généralement bien maîtrisées par ces derniers, se portera plus précisément sur la réflexion avec laquelle ils choisiront de conduire leurs travaux à venir et avec quelle marge de liberté pour eux-mêmes ils oseront le faire.
Ensuite, c'est bien entendu à nos partenaires enseignants de transformer ces approches pédagogiques en fonction de l'évolution des besoins des écoles d'architecture en Chine, besoins dont nous savons tous qu'ils vont évoluer très rapidement et notamment pour l'art et la culture.

Philippe Guérin

东南大学建筑学院
School of Achitecture Southeast University

创立于1927年，是中国现代建筑教育的发源地。学科由建筑学院、建筑研究所、建筑历史遗产保护研究院等组成，并与校城市规划设计研究院、建筑设计研究院形成产学研一体的体制。80年来，建筑学院在学科建设、学术队伍建设、科学研究、人才培养以及国际学术交流等方面都取得了卓越的成绩，为国家培养了近三千名高级建设人才，其中院士6名，建设工程设计大师10名。

Founded in 1927, it's the birthplace of Chinese modern architecture education. It consists of School of Architecture, Institute of Architecture, Architecture History Research Institute and other components. In the past 80 years, it has achieved excellent results and about 3000 senior professionals, including 6 academicians, 10 masters in construction and engineering.

菲利普·葛汉
Philippe Guérin

画家、建筑师、教师，先后在法国多所国立高等建筑学院任教，主要从事"表现艺术及方法"和艺术理论及历史方面的教学工作。近年来，他陆续开始在中国、意大利等地讲学。

A painter and an architect, he has also been teaching at various national graduate schools of architecture in France.
His main areas of teaching include ATR (Arts and Techniques of Representation) as well as art history and art theory. He recently began teaching abroad, in China and Italy.

宋 磊
Song Lei

隽石（中国）文化传播机构总策划，长期关注并投身于文化及教育事业，策划并组织艺术、建筑、设计领域的中外文化及教学交流活动，同时从事出版领域的专业工作。

Director of Junstone Cultural Communication Co. Ltd.
Over the past years, he has been very active in planning and organizing cultural and educational exchanges (involving art, architecture, and design) between China and abroad. Simultaneously, he also works in publishing.

王 盈
Wang Ying

毕业于清华大学建筑学院，法国国家建筑师，法国环境与景观学硕士，法国理想艺术联盟执行理事，长期从事建筑设计工作，同时致力于艺术及设计领域的法中文化及教学交流。

Architect graduated from Tsinghua University in China, Architect DPLG in France, Master of Environment and Landscape, Executive Director of the Association "Alliance IDEART".
Besides his work as an architect, he is also very active in the Sino-French cultural and educational exchanges (art and design).